Advanced Dielectric, Piezoelectric and Ferroelectric Thin Films

Technical Resources

Journal of the American Ceramic Society

www.ceramicjournal.org

With the highest impact factor of any ceramics-specific journal, the *Journal of the American Ceramic Society* is the world's leading source of published research in ceramics and related materials sciences.

Contents include ceramic processing science; electric and dielectic properties; mechanical, thermal and chemical properties; microstructure and phase equilibria; and much more.

Journal of the American Ceramic Society is abstracted/indexed in Chemical Abstracts, Ceramic Abstracts, Cambridge Scientific, ISI's Web of Science, Science Citation Index, Chemistry Citation Index, Materials Science Citation Index, Reaction Citation Index, Current Contents/ Physical, Chemical and Earth Sciences, Current Contents/Engineering, Computing and Technology, plus more.

View abstracts of all content from 1997 through the current issue at no charge at www.ceramicjournal.org. Subscribers receive full-text access to online content.

Published monthly in print and online. Annual subscription runs from January through December. ISSN 0002-7820

International Journal of Applied Ceramic Technology

www.ceramics.org/act

Launched in January 2004, *International Journal of Applied Ceramic Technology* is a must read for engineers, scientists,and companies using or exploring the use of engineered ceramics in product and commercial applications.

Led by an editorial board of experts from industry, government and universities, *International Journal of Applied Ceramic Technology* is a peer-reviewed publication that provides the latest information on fuel cells, nanotechnology, ceramic armor, thermal and environmental barrier coatings, functional materials, ceramic matrix composites, biomaterials, and other cutting-edge topics.

Go to www.ceramics.org/act to see the current issue's table of contents listing state-of-the-art coverage of important topics by internationally recognized leaders.

Published quarterly. Annual subscription runs from January through December. ISSN 1546-542X

American Ceramic Society Bulletin

www.ceramicbulletin.org

The *American Ceramic Society Bulletin*, is a must-read publication devoted to current and emerging developments in materials, manufacturing processes, instrumentation, equipment, and systems impacting the global ceramics and glass industries.

The *Bulletin* is written primarily for key specifiers of products and services: researchers, engineers, other technical personnel and corporate managers involved in the research, development and manufacture of ceramic and glass products. Membership in The American Ceramic Society includes a subscription to the *Bulletin*, including online access.

Published monthly in print and online, the December issue includes the annual *ceramicSOURCE* company directory and buyer's guide. ISSN 0002-7812

Ceramic Engineering and Science Proceedings (CESP)

www.ceramics.org/cesp

Practical and effective solutions for manufacturing and processing issues are offered by industry experts. CESP includes five issues per year: Glass Problems, Whitewares & Materials, Advanced Ceramics and Composites, Porcelain Enamel. Annual subscription runs from January to December. ISSN 0196-6219

ACerS-NIST Phase Equilibria Diagrams CD-ROM Database Version 3.0

www.ceramics.org/phasecd

The ACerS-NIST Phase Equilibria Diagrams CD-ROM Database Version 3.0 contains more than 19,000 diagrams previously published in 20 phase volumes produced as part of the ACerS-NIST Phase Equilibria Diagrams Program: Volumes I through XIII; Annuals 91, 92 and 93; High Tc Superconductors I & II; Zirconium & Zirconia Systems; and Electronic Ceramics I. The CD-ROM includes full commentaries and interactive capabilities.

Advanced Dielectric, Piezoelectric and Ferroelectric Thin Films

Ceramic Transactions Volume 162

Proceedings of the 106th Annual Meeting of The American Ceramic Society, Indianapolis, Indiana, USA (2004)

Editors
Bruce A. Tuttle
Chonglin Chen
Quanxi Jia
R. Ramesh

Published by
The American Ceramic Society
PO Box 6136
Westerville, Ohio 43086-6136
www.ceramics.org

Advanced Dielectric, Piezoelectric and Ferroelectric Thin Films

For information on ordering titles published by The American Ceramic Society, or to request a publications catalog, please call 614-794-5890, or visit our website at www.ceramics.org

ISBN 1-57498-183-8

Contents

Preface

This Ceramic Transactions volume contains selected papers from the Advanced Dielectric, Piezoelectric and Ferroelectric Thin Films Symposium held at The American Ceramic Society's 106th Annual Meeting, April 18-21, 2004 in Indianapolis, Indiana. This symposium brought together scientists and engineers to present the latest advances concerning the synthesis and characterization of dielectric, piezoelectric and ferroelectric thin films. Dielectric, piezoelectric and ferroelectric thin films have a tremendous impact on a variety of commercial and military systems including tunable microwave devices, memories, MEMS devices, actuators, and sensors. Recent work on piezoelectric characterization, AFE to FE dielectric phase transformation dielectrics, solution and vapor deposited thin films were among the topics presented. As such, chemical preparation of thin films, structure-property relationships, and process integration issues were among the topics highlighted. Novel approaches to nano-structuring, characterization of material properties and physical responses at the nanoscale were also emphasized.

We would like to thank all of the participants in the symposium and especially those who contributed to this volume. Many thanks are also due to the staff of The American Ceramic Society for their assistance in handling a wide variety of issues before, during, and after the meeting and for helping to produce this proceeding.

Bruce A. Tuttle
Chonglin Chen
Quanxi Jia
R. Ramesh

CHEMICAL SOLUTION DEPOSITION OF $CaCu_3Ti_4O_{12}$ THIN FILMS

Mark D. Losego and Jon-Paul Maria
North Carolina State University
1001 Capability Drive, Raleigh, NC 27695

ABSTRACT

A chemical solution deposition method is used to prepare calcium copper titanate thin films on platinized silicon substrates. The impact of annealing temperature and stoichiometry on phase formation and dielectric properties is investigated. Through x-ray diffraction analysis, an intermediary phase, identified here as Cu_3TiO_4, is shown to emerge before the crystallization of $CaCu_3Ti_4O_{12}$ at 725°C. The temperature at which conversion occurred was mildly dependent on copper stoichiometry. Permittivities between 200 and 400 were observed for all cases; these values are significantly smaller than others reported recently. Field dependent measurements show a voltage variable permittivity which is linked to mobile charges in the CCT crystals and Schottky barriers at both electrode interfaces.

INTRODUCTION

For more than twenty-five years, calcium copper titanate (CCT) and its related isomorphic ($AB_3M_4O_{12}$) compounds have been known to exhibit a perovskite-like structure. However, because the MO_6 octahedra in this structure are tilted[1], the unit cell becomes a centrosymmetric supercell of 8 perovskite-like units. Unlike common ferroelectric perovskites like barium titanate and lead zirconate titanate, that are well known for their high dielectric permittivities, materials with the CCT structure are crystallographically forbidden from exhibiting ferroelectricity, and, until recently, dielectric measurements were not even reported.

In 2000, Subramanian et al. reported for the first time on the "giant" dielectric constant observable in CCT.[2] Since that discovery, dielectric constants of 80,000 for bulk single crystals[3] and 10,000 for bulk polycrystalline samples[2,3] have been reported at test frequencies of 10 kHz. Epitaxial CCT films grown by pulsed-laser deposition (PLD) have been reported to display relative permittivities of 10,000 at 1 MHz.[4] Further work revealed that although other $AB_3M_4O_{12}$ isomorphs do not show permittivities as high as CCT, they do exhibit dielectric constants that are 10 to 50 times higher than predicted by the Clausius-Mosotti equation.[5]

A complete explanation for this behavior has not yet been identified. Most current models suggest a large extrinsic contribution. For instance, an internal barrier layer mechanism was proposed by Sinclair, based on impedance spectroscopy analysis of polycrystalline bulk CCT pellets.[6,7] However, classical internal barrier layer capacitors rely on the dielectric contributions of non-conducting grain boundaries and as such are inconsistent with the results reported for single-crystal CCT. One proposed explanation is that twinning, which appears to be densely present in bulk samples, may act as the internal barrier.[5,8]

This paper will discuss a chemical solution deposition (CSD) method for preparing CCT thin films. To date, all CCT films reported in the open literature have been deposited using PLD and most of these have exhibitied epitaxial morphologies. Synthesizing films by CSD however, offers an important advantage for investigating the origin of giant permittivity in CCT: The stoichiometry can be controlled with the precision of bulk synthesis methods, while the thin layer geometry affords easy measurement of field dependent dielectric properties and electrical transport. Finally, CSD deposition represents a reasonable method to cost-effectively mass-produce commercial CCT thin films.

EXPERIMENTAL DETAILS

Methanol-based, chelated solutions were prepared using calcium acetate hydrate (Aldrich, 99.99% pure, <6 mol% H_2O), copper (II) acetate (Aldrich, 98% pure), and titanium (IV) isopropoxide (Alfa-Aesar, >98% pure) as cation sources. Solutions containing each metallic species were prepared separately and then combined by mass to maintain stoichiometric accuracy. During preparation solutions were maintained at 85°C and continually stirred at 300 rpm. Final CCT solutions had concentrations of about 0.055 M and appear to be stable for between 3 and 6 weeks at room temperature. To evaluate the effects of composition, off stoichiometric solutions were also prepared, including $Ca_{0.26}Cu_{0.74}TiO_4$ (Ca-rich); $Ca_{0.24}Cu_{0.76}TiO_4$ (Cu-rich); $Ca_{0.25}Cu_{0.75}Ti_{0.99}O_4$ (Ti-deficient); $Ca_{0.25}Cu_{0.75}Ti_{1.01}O_4$ (Ti-rich). Difficulties in ascertaining the exact amount of hydration in the calcium source may have led to solutions containing less calcium than expected. However, this error is believed to be less than 1 mol%, thereby maintaining the validity of compositional property trends.

Solutions were deposited through a 0.2 μm filter onto platinized silicon substrates. Films were spun at 3000 rpm for 30 s and then dried on a hotplate at 250°C for 5 min. Pyrolysis was carried out in an open air furnace at 400°C for 15 min. These steps were repeated to reach the desired thickness. Crystallization anneals were performed between 500°C and 800°C. Samples prepared for electrical characterization contained four deposition layers and were crystallized at 725°C and 750°C. Platinum top electrodes were dc magnetron sputtered on these crystallized samples at 30 mTorr Ar, 250 W for 30 s.

Crystallization behavior and phase development in these films was studied with a Bruker AXS D-5000 x-ray diffractometer equipped with a HighStar® area detector. Grain size analysis was performed using the lineal intercept procedure on images taken with a Digital Instruments Dimension 3100 atomic force microscope (AFM). AFM scans were taken over a 1 μm^2 area in tapping mode. Film thickness was ascertained from cross-sectional images taken with a Hitachi S-3200 scanning electron microscope. Capacitance and dielectric loss data was collected with an HP 4192A impedance analyzer. Leakage current information was obtained using a Keithley 617 Electrometer.

CRYSTALLIZATION BEHAVIOR

X-ray diffraction (XRD) scans demonstrating the phase behavior during crystallization of stoichiometric CCT films are shown in figure 1. The intensities of the CCT reflections shown in these scans are indicative of randomly oriented polycrystalline CCT. Furthermore, these scans clearly indicate the presence of a low temperature intermediary phase prior to the crystallization of CCT. The location of this peak matches well with the *101* reflection for Cu_3TiO_4 (PDF Card #27-199). Figure 2 summarizes the phase behavior data collected from XRD scans performed on non-stoichiometric CCT films. This data reveals that the intermediary phase has the shortest lifetime in the calcium rich sample. Since the calcium rich sample would lessen the amount of

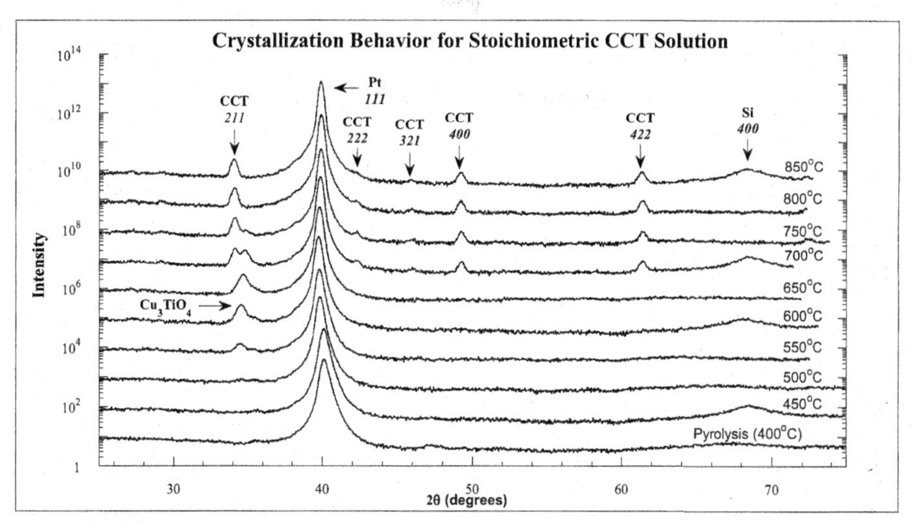

Figure 1: X-ray diffraction data collected for CSD deposited stoichiometric CCT films fired between 400°C and 850°C

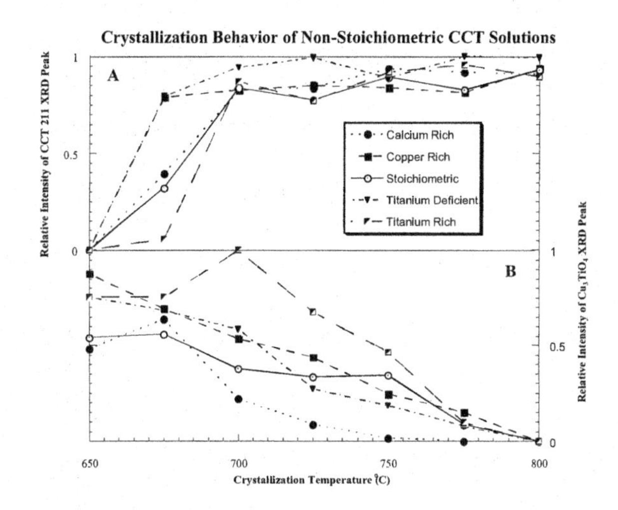

Figure 2: Relative XRD peak intensities for (A) the 211 CCT reflection and (B) the Cu_3TiO_4 in CCT films of varying composition crystallized over a range of temperatures

copper available for forming Cu_3TiO_4, this observation appears to be consistent with the proposed intermediary phase.

The electrical characterization data presented later in this paper was collected from samples crystallized at 725°C. At this temperature, all samples still contain some remaining intermediary phase, but this phase appears to have the highest volume fraction in the titanium rich and copper rich samples.

MICROSTRUCTURAL ANALYSIS

Scanning electron microscopy performed on cross-sections of CCT films (figure 3) reveals an isometric grain morphology. Grain size is determined with atomic force microscopy (see figure 4). The 0.3 μm thick (4 layer) sample has an average grain size of 67 nm and an RMS roughness of 4.4 nm while the 0.5 μm thick (7 layers) sample had an average grain size of 76 nm and an RMS roughness of 5.2 nm. These grain sizes are believed to be the smallest reported for CCT. As such, given the importance of microstructural features to determine the dielectric response of barrier layer capacitors, these small sizes should be pertinent in future interpretation.

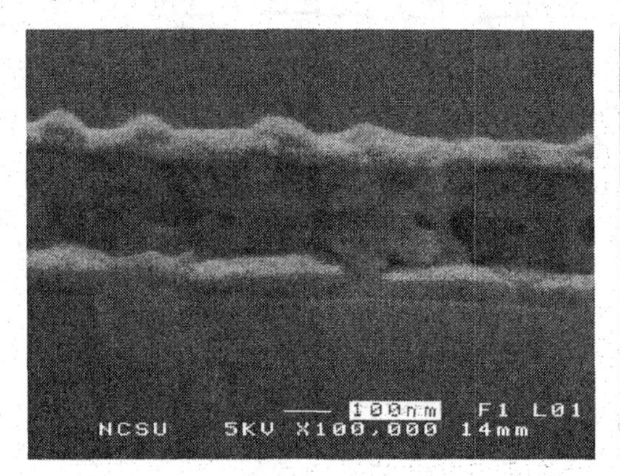

Figure 3: Cross-sectional image of CCT film taken with a scanning electron microscope.

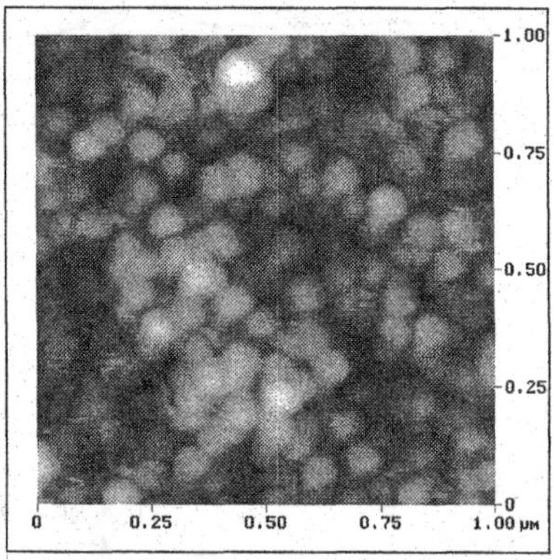

Figure 4: AFM image taken of the 0.3 μm thick CCT film surface.

ELECTRICAL RESULTS

The electrical data shown in Figures 5 and 6 is collected from CCT films crystallized at 725°C. Most of the samples processed in parallel with these films and then annealed at 750°C were extremely conductive with an impedance phase angle of approximately zero at 10 kHz. Similar results were observed in samples that were less than 4 layers thick (0.3 μm). Only the titanium-rich and titanium-deficient samples exhibited dielectric behavior within the examined frequency range. The loss tangent was consistently lower for these 750°C samples. The difference was approximately 25% at 10 kHz. An explanation for why most of these samples were conductive is ongoing and may indicate the necessity for a post-anneal oxidation treatment, which has been shown to lower conductivity in PLD deposited CCT films.[9]

As illustrated in figures 5A and 6A, the permittivity of these CSD derived CCT films is 4 to 10 times lower than values recently reported in the literature for PLD deposited films.[10,11,12] This could be due to several factors including the residual intermediary phase or the fine grained microstructure.

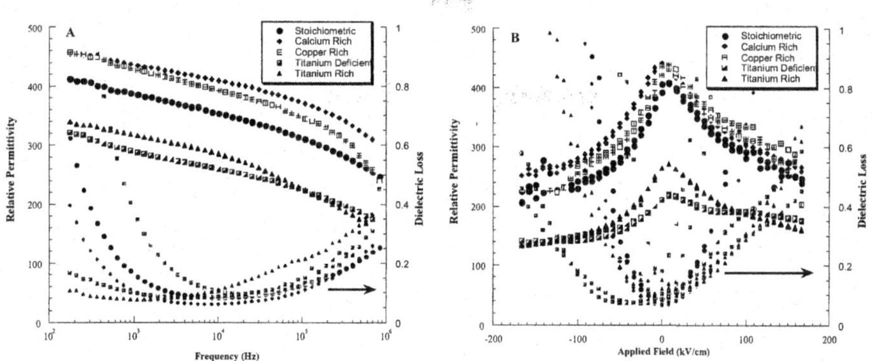

Figure 5: Dielectric response vs. (A) frequency and (B) applied electric field for CCT films with varying composition. Electric field dependent data was collected at 10 kHz.

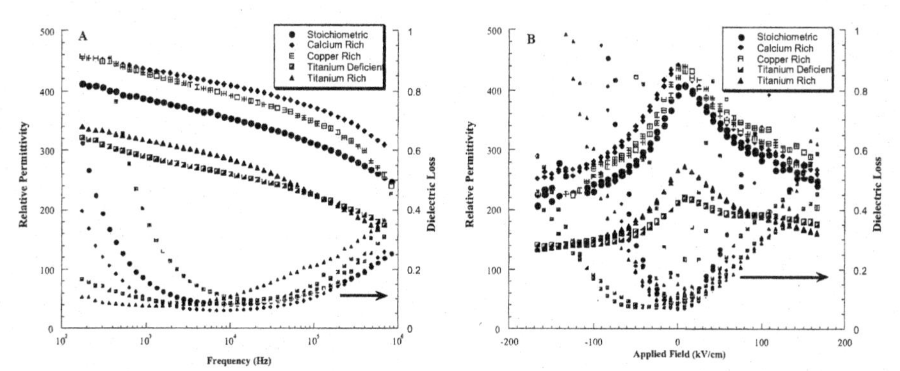

Figure 6: Dielectric response vs. (A) frequency and (B) applied electric field for CCT films of varying thickness. Electric field dependent data was collected at 10 kHz.

XRD data indicates that the 7 layer sample contains less residual intermediary phase than the 8 layer sample, thereby supporting the necessity of complete transformation for optimal dielectric response. Figure 5A also appears to indicate that maintaining the proper divalent to tetravalent cation ratio is necessary for achieving higher permittivities.

However, these films do exhibit tunability with applied bias as shown in figures 5B and 6B. Such a response has not been reported previously for CCT films. Compared to ferroelectric materials, where dielectric loss peaks near zero bias due to domain switching, the tan δ in these CCT films increases with increasing bias. This increase in conductivity with bias may result from the redistribution of charge carriers to a film/electrode interface, and a subsequent reduction in the Schottky barrier. This effect is also likely to be responsible for the apparent dielectric tuning.

Leakage current data is plotted in figure 7. Thicker films exhibit lower conductivity which is consistent with the hypothesis that most of the dielectric loss is due to mobile charge carriers.

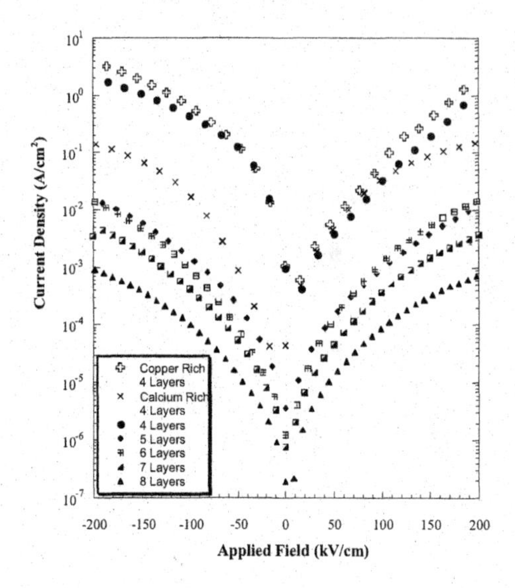

Figure 7: Leakage current data for CCT films of varying thickness and composition.

The addition of calcium also appears to lower DC losses; this result may reflect the fact that the calcium rich sample contained less of the intermediary phase.

CONCLUSIONS

Chemical solution deposition has proven to be a possible means for preparing calcium copper titanate thin films. Currently, however, the dielectric constant of these films is significantly lower than the ultra-high values reported by several authors. Since the ultra-high permittivities are believed to result from internal barriers, like crystallographic twins, we conclude that such boundaries are not present in the current CSD films. The reason for this absence is not understood, however, possibilities include very fine grain size and low process temperatures (as compared to bulk ceramic methods). We show that dielectric loss and leakage current can be improved with increased film thickness and possibly the addition of excess calcium, however, these changes likely correspond to the intrinsic permittivity of the CCT lattice and not the ultra-high permittivity associated with the barrier-layer mechanism.

This material is based upon work supported under a National Science Foundation Graduate Research Fellowship

REFERENCES

[1]B. Bochu et al. "Synthèse et Caractèrisation d'une Sèrie de Titanates Pèowskites Isotypes de $[CaCu_3](Mn_4)O_{12}$," J. *Solid State Chem.* **29** 291-298 (1979).

[2]M. A. Subramanian et al. "High Dielectric Constant in $ACu_3Ti_4O_{12}$ and $ACu_3Ti_3FeO_{12}$ Phases," *J. Solid State Chem.* **151** 323-325 (2000).

[3]C. C. Homes et al. "Optical Response of High-Dielectric-Constant Perovskite-Related Oxide," *Science* **293** 673-676 (2001).
[4]Y. Lin et al. "Epitaxial Growth of Dielectric $CaCu_3Ti_4O_{12}$ Thin Films on (001) $LaAlO_3$ by Pulsed Laser Deposition," *Appl. Phys. Let.* **81** [4] 631-633 (2002)
[5]M A. Subramanian and A. W. Sleight, "$ACu_3Ti_4O_{12}$ and $ACu_3Ru_4O_{12}$ Perovskites: High Dielectric Constants and Valence Degeneracy," *Solid State Sciences* **4** 347-351 (2002)
[6]D. C. Sinclair et al. "$CaCu_3Ti_4O_{12}$: One-Step Internal Barrier Layer Capacitor," *Appl. Phys. Let.* **80** [12] 2153-2155 (2002).
[7]T. B. Adams, D. C. Sinclair, and A. R. West, "Giant Barrier Layer Capacitance Effects in $CaCu_3Ti_4O_{12}$ Ceramics," *Adv. Mater.* **14** [18] 1321-1323 (2002).
[8]M. H. Cohen et al. "Extrinsic Models for the Dielectric Response of $CaCu_3Ti_4O_{12}$," *J. Appl. Phys.* **94** [5] 3299-3306 (2003).
[9]Tselev et al. "Intrinsic Dielectric Response of the Colossal Dielectric Constant Material $CaCu_3Ti_4O_{12}$," *Condensed Matter,* under review.
[10]L. Fang and M. Shen, "Deposition and Dielectric Properties of $CaCu_3Ti_4O_{12}$ Thin Films on Pt/Ti/SiO_2/Si Substrates using Pulsed-Laser Deposition," *Thin Solid Films*, **440** 60-65 (2003).
[11]W. Si et al. "Epitaxial Thin Films on the Giant-Dielectric-Constant Material $CaCu_3Ti_4O_{12}$ Grown by Pulsed-Laser Deposition," *Appl. Phys. Let.* **81** [11] 2056-2058 (2002).
[12]Y. L. Zhao et al. "High Dielectric Constant in $CaCu_3Ti_4O_{12}$ Thin Films Prepared by Pulsed Laser Deposition," *Thin Solid Films* **445** [1] 7-13 (2003).

THE TEMPORAL EFFECTS IN DC-BIASED $PbNb(Zr,Sn,Ti)O_3$ ANTIFERROELECTRIC THIN FILMS

Jiwei Zhai, Haydn Chen
Department of Physics and Materials Science
City University of Hong Kong
Kowloon, Hong Kong, China

Eugene V. Colla
Department of Physics
University of Illinois at Urbana-Champaign
Urbana, IL 61801, USA

ABSTRACT

The effects of the history of the DC bias application on the polarization of the antiferroelectric $Pb_{0.99}Nb_{0.02}(Zr_{0.82}Sn_{0.12}Ti_{0.04})_{0.98}O_3$ (PNZST) thin films were studied at room temperature. It was shown that in these films, grown by a sol-gel process, the antiferroelectric ordering was temporarily destroyed in time scale of approximately one second or less by application of a DC electrical field bias along the surface normal direction. After removing the DC bias the film relaxed slowly back to the initial antiferroelectric state with a relaxation time of approximately a few thousands of seconds. This temporal phenomenon was dependent on film thickness. We attributed these behaviors to the accumulation of space charges near the film/substrate interface region under DC bias to set up a self-polarization, followed by a relaxation upon removal of DC bias by way of migration of space charges via ionic diffusion.

The antiferroelectric (AFE) materials can be characterized with a zero net polarization because of the alternating alignment of local dipoles. The application of an external electrical field with sufficiently high magnitude to an AFE specimen could cause the flipping of dipoles, resulting in their parallel alignment and the appearance of a macroscopic polarization similar to the ferroelectric (FE) materials. In some materials (i.e. soft antiferroelectric) the balance between ferroelectric and antiferroelectric orderings is rather fragile so that the AFE-FE switching becomes easier by the application of some external forces (e.g. electrical field [1], or hydrostatic pressure [2]). In the case of thin films this switching behavior could depend on the film/substrate interface properties (strain, space charge, etc). For example, the substrate/film interface charge can produce self-polarization of the film even without the presence of an external electrical field [3,4].

Thin films of antiferroelectric materials are of interest because of their applications in microelectronic devices; understandably they must possess acceptable performance to ensure the long-term reliability. The polarization characteristics of the antiferroelectric $PbZrO_3$ (PZ), $PbLa(Zr,Sn,Ti)O_3$ (PLZST), and $PbNb(Zr,Sn,Ti)O_3$ (PNZST) [5-10] thin films under DC bias have not been studied in any sufficient details. In this letter, we present the analysis of the polarization switching in PNZST thin films as a function of DC bias, AC frequency and relaxation time.

The $Pb_{0.99}Nb_{0.02}(Zr_{0.82}Sn_{0.12}Ti_{0.04})_{0.98}O_3$ (PNZST) antiferroelectric thin films of different thickness were prepared by a sol-gel process [11]. Thin films were deposited by spin coating at 3000 rpm for 30s on the $LaNiO_3$ (LNO) buffered $Pt/Ti/SiO_2/Si(100)$ substrates. The thickness of $LaNiO_3$, Pt, Ti, and SiO_2 buffer layers on Si substrates were 150 nm, 150 nm, 50 nm, and 150 nm, respectively. The LNO layer serves two purposes: one as a conducting oxide bottom electrode and the second as a template layer promoting PNZST growth with highly (100) preferred orientation [11]. Each spin-on PNZST layer was heat treated at 500°C for 5 minutes. The spin coating and heat-treatment processes were repeated several times to obtain the expected film thickness. To form the pure perovskite phase a capping layer of 0.8 M PbO precursor solutions was added at the top of the film before the final annealing at 700°C for 30 minutes. For the electrical measurements the top gold electrodes of $200\times200\ \mu m^2$ were deposited by DC sputtering.

Electrical properties were studied using an Agilent 4284A LCR meter to measure the capacitance (C) vs. voltage (V), and a Radiant Technologies Precision Pro ferroelectric tester to measure the polarization (P) vs. field (E). The phase structure of the PNZST was studied by X-ray diffraction using a SIEMENS D-500 powder X-ray diffractometer with a filtered CuK_α radiation. The thickness of film was measured using scanning electron microscopy (SEM).

The electrical field dependence of the dielectric constant (ε') obtained from the PNZST films (on LNO-buffered Si substrates) of various thicknesses is shown in Fig. 1. The dielectric constant was calculated from the measured capacitance along with the measured film thickness and the top electrode area. A bipolar triangular voltage of 650s period was applied to the sample as bias field (E), and the LCR meter was used for capacitance measurements. Each panel of graph in Fig. 1 represents two consecutive cycles of ε'-E measurements. The solid line depicts the first cycle, while the second cycle data by the dashed line. All measurements presented in Fig. 1 were made by starting with the positive bias field to a level that the AFE-FE switching reached completion, termed "poling". Thereafter the DC bias was reversed towards the negative field. Consequently two peaks are present in the positive half period of the scan signifying the AFE-FE forward switching and the FE-AFE back switching. Two more peaks of similar nature appear in the negative half period of the scan for thicker films.

It is clear that there is no big difference between the first and the second curves for 730 nm thick film (Fig. 1(d)), but with decreasing film thickness the second scan curves show diminishing AFE features as the four peaks begin to get smeared and eventually become unnoticeable. The FE-AFE back switching becomes less and less pronounced and in the case of 170 nm film (Fig. 1(a)) the second curve does not show the four peaks typical of the forward AFE-FE and the backward FE-AFE switching for both positive and negative DC bias. As a matter of fact the third and the subsequent scans show nearly identical profiles to the second curve. It is concluded from Figures 1(a) to 1(b) that at this relatively slow measuring period

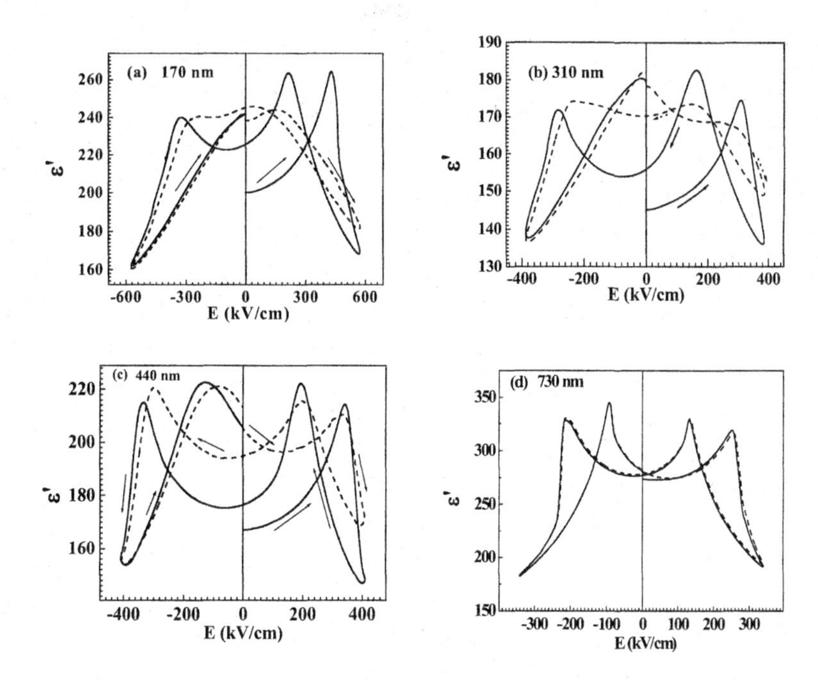

Fig.1 Dielectric constant versus electrical field (ε'-E) characteristics of PNZST thin films deposited on LNO substrates with thickness of (a) 170 nm, (b) 310 nm, (c) 440 nm, and (d) 730 nm. Dielectric constants were obtained from the measured capacitance values. The measurement time for each cycle was 650s. Solid lines represent data from the first cycle, while the dashed lines are for the second cycle measurements.

(650s) and for sufficiently thin films (< 730 nm), once the FE state is established the return to the AFE order is retarded or even destroyed. Excursion first to the saturation bias destroys the AFE order but creates an FE order, which does not disappear even after removing the field (no second peak was present while the experiment started with positive bias).

Similar conclusion was reached from a different set of experiments where the P-E (polarization vs. electrical field) relationships were measured under fast time (1 millisecond) period and after application of a DC bias. Shown in Fig. 2 are two P-E hysteresis loops observed for the 440nm thick PNZST films. The first curve (solid line) corresponds to the fast measurement (1 millisecond) done on the virgin sample. The second P-E curve (dashed line) was taken after the initial C-V experiment was completed and the sample was exposed to the DC electrical field higher than the E_{AFE-FE} switching voltage for more than 10 min time. Comparing Fig. 2 with Fig. 1(c) for the same film thickness, one notices that during the fist measurement cycle although the P-E (Fig. 2) curve shows a double hysteresis loops, the C-V curve (Fig. 1(c)) clearly signifies the presence of the FE order as indicated by the asymmetry of the four peaks and the progressive increase of capacitance at zero volt. From this comparison it can be concluded that the time (τ_1) necessary to destroy the AFE ordering by the electrical field is no more than the order of seconds. The hysteresis loops shown in Fig. 2 were obtained with the scanning time of 1 millisecond and all experimental observations confirm that this time is too short to break the AFE alignment in the sample. But after the initial C-V test (completed in 650s) where poling has taken effect, AFE order is obviously destroyed to the extent that the subsequent P-E curve does not show the characteristic double hysteresis loops.

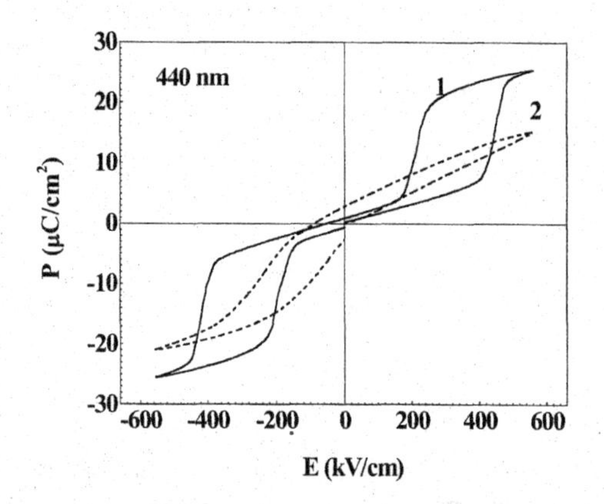

Fig. 2 Hysteresis loops of a 440 nm thick PNZST thin film deposited on LNO/Pt/Si/SiO_2/Si in (a) the virgin condition and (b) after electrical field application longer than the time duration τ_1 (~1s). The measurement time for each P-E cycle was 1 millisecond.

The stability of the poled state can be checked by studying the relaxation behavior from the poled FE state back to AFE ordering; this has been accomplished in Fig. 3 where the "positive" branches of the double hysteresis loops taken in different time intervals after the first full cycle is displayed. (For the P-E measurements, approximately 1 millisecond was used for each time interval, which was faster enough to not cause any further deteriation of AFE or FE order at room temperature.) It was found that the electrical field-induced FE state is metastable and it relaxes back to the original AFE phase over time. Characteristic time (τ_2) of this relaxation is much longer than τ_1. For reference, curve 0 corresponds to the virgin sample (not exposed to negative bias) curves 1-5 represent the evolution of the hysteresis loop after the sample was subjected to negative bias for the time longer than τ_1. In the insert to Fig 3 the time dependence of residual polarization (i.e. P(E=0, t)) is shown. The horizontal axis marks the moments when the loops 1-5 were recorded. From Fig. 3 it is evident that the sample eventually returns to its original state.

The total recovery time exceeds 10^5 s at room temperature for this particular film thickness (440nm). The dashed line in the insert to Fig. 3 corresponds to the initial remanent polarization of the unpoled sample. We also found that the sample can be recovered much faster by annealing it at temperature over the Curie temperature (~220^0C) – the temperature above which the system becomes paraelectric.

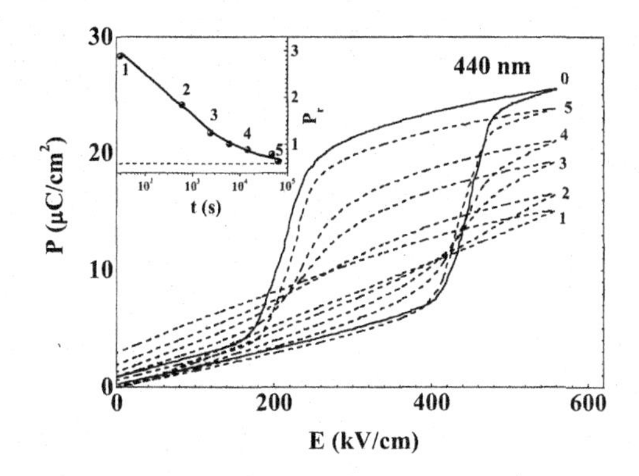

Fig. 3 Relaxation to the original AFE state after the initial DC field poling is observed by the restoration of hysteresis loop on the P-E curves. The solid line represents measurements done on the virgin sample, whereas dashed lines record data taken at different times after the initial poling cycle. In the insert, the time dependence of the field induced remanent polarization is shown.

We believe that the observed phenomenon of the temporary destruction of the AFE state in PNZST films by the DC field application is linked to the properties at the film/substrate interface layer. This comes from the thickness dependence of the stability of the AFE phase after field application. The thicker the films are, where the contribution of the interface layer becomes less significant, the less drastic are the property changes after initial poling.

In order to explain the above experimental results, we employed a simple diffusion model to describe the relaxation kinetics in the film. From Fig. 2 it is clear that before the slow negative bias was applied to the films the observed hysteresis loop was quite symmetric; this means that it is unlikely that there exist strong intrinsic fields responsible for self-polarization [3,4]. On the other hand after the application of negative field the P-E curve became asymmetric (curve 2 in Fig. 2). From curve 1 in Fig. 2 it is evident that after AFE-FE switching the high polarization phase is stable until the electrical field is over ~200 kV/cm. It means that if after the field application the film loses the AFE features, we can expect the existence of an induced electrical field with comparable magnitude. The saturation polarization in this case (curve 2 in Fig. 2) is smaller than that in virgin sample tested by fast field application (curve 1 in Fig. 2) and the new state is expected to be a partially ordered FE in a strong induced field. Indeed, in the case of the negative bias (the external applied field is in parallel with the induced one) the saturation polarization is bigger and the loop shape (negative branch of curve 2 in Fig. 2) resembles the typical FE hysteresis curve. For positive branch the external field works against the induced intrinsic polarization and the loop shape is much slimmer and not saturated. Both observed processes (AFE phase destruction by the electrical field and its recovery after relaxation) could be characterized by two time constants (τ_1 ~1s and τ_2 ~ 10^3s with τ_2/τ_1 being ~10^3). These temporal effects could be explained in terms of ion diffusion through the entire film, first to set up the space charge accumulation in the interface regions under a DC bias, then to relax without the bias. Obviously this interfacial effect is thickness dependent; it becomes less significant for thicker films.

If D is the diffusivity and L the diffusion distance (in our case L equals to the film thickness), τ_2 (representing the recovery time by way of ionic charge diffusion after the destruction of the AFE state due to poling) can be expressed as

$$\tau_2 \approx \frac{L^2}{D} \qquad (1)$$

For diffusion induced by the electrical field, E, which causes the movement of ionic charges to the interfaces to set up a reverse polarization, the corresponding time period τ_1 may be estimated as $\tau_1 \sim \frac{L}{V_f}$, in which V_f is the particles flow velocity that can be estimated from the Einstein relation between the diffusivity and mobility [12]: $V_f \approx Eq\mu$, here q is the diffusing particle charge and μ the mobility. Taking into account that $\mu = \frac{D}{kT}$ we can find the following expression for τ_1

$$\tau_1 \approx \frac{L \cdot kT}{EqD}. \qquad (2)$$

Replacing $E = V_b/L$, where V_b is the bias voltage, and substituting Eq. (1) into Eq. (2), we obtain

$$\tau_1 \approx \frac{L^2 \cdot kT}{qDV_b} \approx \frac{\tau_2 \cdot kT}{qV_b}. \qquad (3)$$

Using realistic values for V_b~30V and kT~25meV (room temperature) even for q=1 the ratio $\frac{\tau_2}{\tau_1}$ ~1200; this is consistent with the observed value. Consequently, the diffusion driven relaxation effect is a plausible explanation to the observed temporal effect on the relaxation property from the poled FE state back to the stable AFE state.

In summary, the AFE alignment of PNZST thin films can be destroyed by the application of the DC electrical field for a sufficiently long time (e.g. >1s at room temperature). The effect can be easily revealed by a drastic change in the measured P-E hysteresis loops as a function of time. Short time field application (~1 ms or less) does not affect AFE properties of the film, however. The induced FE order resulted from field application is not stable so it will relax back to the original AFE state with a characteristic time of a few thousands of seconds at room temperature. This temporal phenomenon is dependent on the film thickness, which implies that the film/substrate interface is responsible for the observed effects. We have used the ionic diffusion model to explain the relaxation effects. The appearance of the induced intrinsic electrical fields due to space charge accumulation at the interfaces, which destroys the proximate AFE ordering, occurs after poling; this is followed by ionic diffusion to relax back to the original AFE state when the bias field is removed. These space charges at interfaces would then affect the dielectric properties, giving rise to the observed temporal effects. These effects are obviously thickness dependent as the interfacial effect becomes less dominant for thicker films.

Acknowledgments

This research was fully supported by grants from the City University of Hong Kong under the project numbers of 7001464 and 9380015. We are gratefully to Prof. M. B. Weissman of University of Illinois at Urbana-Champaign for fruitful discussions. The technical assistance of Daniel Yau is acknowledged.

References

[1] Wai-Hung Chan, Haydn Chen and EugeneV. Colla, Appl. Phy. Let. (2003) In print.
[2] Zhuo Xu, Yujen Feng, Shuguang Zheng, An Jin, Fanglin Wang, and Xi Yao, J. Appl. Phys., **92**, 2663, (2002)
[3] I. P. Pronin, E. Yu. Kaptelov, E. A. Tarakanov, T. A. Shaplygina, V. P. Afanasjev, and A. V. Pankrashkin, Physics of the Solid State, **44**, 769 (2002).
[4] V. P. Afanasjev, A. A. Petrov, I. P. Pronin, E. A. Tarakanov, E. Ju Kaptelov and J. Graul, J. Phys.: Condens. Matter **13,** 8755 (2001).
[5] K. Yamakawa, S. Trolier-Mckinstry, and J. P. Dougherty, S. B. Krupanidhi, Appl. Phys. Lett. **67,** 2014 (1995).
[6] R.Seveno, H. W. Gundel, and S. Seifert, Appl. Phys. Lett. **79,** 4204 (2001).
[7] Baomin Xu, Paul Moses, Neelesh G. Pai, and L. Eric Cross, Appl. Phys. Lett. **72,** 593 (1998).
[8] Jae Hyuk Jang, Ki Hyun Yoon, and Hyun Jung Shin, Appl. Phys. Lett. **73**(1998) 1823.
[9] Baomin Xu, Yaohong Ye, Qing-Ming Wang, and L. Eric Cross, J. Appl. Phys. **85,** 3753 (1999).
[10] S. S. Sengupta, D. Robert, J. F. Li, M. C. Kim, D. A. Payne, J. Appl. Phys. **78**, 1171 (1995).
[11] Jiwei Zhai, M. H. Cheung, Zheng Kui Xu, Xin Li and Haydn Chen, Appl. Phys. Lett. **81**, 3621(2002).
[12] C. Kittel, H. Kroemer, Thermal Physics, W.H. Freeman and Company, San Francisco (1980).

HIGH ENERGY DENSITY PLZT THIN FILM CAPACITORS

Bruce A. Tuttle, Geoffrey Brennecka, David P. Williams, Mark A. Rodriguez,
Thomas J. Headley and Jill S. Wheeler
Sandia National Laboratories
Albuquerque, NM 87185

ABSTRACT

Development of next generation electronics for MEMS and RF power systems requires miniaturization and integration of high energy density ceramic dielectrics with Si CMOS and MEMS technologies. In this study, high Zr content (70 mol% Zr) lead lanthanum zirconate titanate (HZPLZT) film compositions were evaluated to reduce overall electrical system volume by reducing the size of what is often the largest system component - the capacitors. Approximately twice the energy density was achieved for the HZPLZT films compared to more conventional near morphotropic phase boundary lead zirconate titanate (PZT) based films. Our study considered seven different process parameters and determined that pyrolysis temperature and use of PbO overcoat layers were especially significant in obtaining single phase perovskite layers that provided the highest energy density films. Transmission electron microscopy analyses were essential in the determination of process-property relationships. TEM analyses indicated that the presence of fluorite phase was responsible for PLZT 12/70/30 films with dielectric constants less than 1000. HZPLZT film capacitors were fabricated that exhibit breakdown strengths in excess of 1.4 MV/cm and an energy density of 29 J/cm^3.

INTRODUCTION

High speed decoupling capacitors [1,2] for high density interconnect technologies, new generation pulse discharge capacitors for MEMS applications [3], and RF embedded capacitors for wireless devices are among the applications [4,5] that can be enabled by lead zirconate titanate (PZT) based thin film technology. Dimos, Al-Shareef and coworkers [1,2] have developed decoupling capacitors in conjunction with high density interconnect polymer – Cu technology. Use of donor dopants, such as Nb and La were shown to reduce film leakage current and enhance high temperature breakdown strength by a factor of three compared to undoped near morphotropic (40 mol% < Zr < 60 mol%) PZT films. An important breakthrough for commercial development of thin film capacitors was the ability to achieve greater single layer thickness deposition by Liu and coworkers [6]. More recently, work by Maria and Trolier-McKinstry [4,5] showed that capacitors based on PZT near morphotropic phase boundary compositions (NMPB) could be deposited on foil and embedded in substrates for wireless applications. Kaufman and coworkers [7] have investigated near morphotropic phase boundary compositions with high La contents (approximately 8 mol%) for foil capacitors.

A critical microstructural issue for high energy density PZT based capacitors is the formation of single phase perovskite films. Formation of single phase perovskite films becomes more challenging as Zr content, PbO volatilization, and La content increase. Several research groups [8 - 10] have shown that PZT based film phase evolution proceeds from amorphous to fluorite/pyrochlore to perovskite phases with increasing process temperature. The perovskite phase often nucleates from the bottom electrode and grows in columnar grains through the PZT film thickness. While the pyrochlore or fluorite phase is very tolerant to changes in Pb stoichiometry, the perovskite phase is not. Extrapolated tolerances of Pb stoichiometry, [11] for which the perovskite phase is stable for PZT films processed at 650°C are in the ± 0.1 mol% range. The extrapolation is from Knudsen effusion measurements of bulk PZT ceramics above 1000°C. Thus, if the top of the PZT film becomes Pb deficient (< 50 mol% Pb), a low dielectric constant, lead deficient pyrochlore/fluorite phase (K = 50) forms and will not transform to the perovskite phase. Dilution of the dielectric constant of the film can be severe, especially if planar fluorite layers are formed. Recent work by M. Mandeljc, Kosec and coworkers [12] for PLZT 9.5/65/35 films indicated that PbO content was the critical factor for crystallization of the perovskite phase, even for process temperatures below 500°C. Their Pb deficient films often showed formation of a continuous layer of fluorite phase at the top of the PLZT films with a columnar perovskite phase present below the fluorite phase.

The film composition selected for our present study was PLZT 12/70/30. There were several reasons to investigate this film composition. First, compared to near morphotropic and PLZT 9/65/35 films there is greater dielectric constant linearity with field. This is a very important property for enhancement of energy storage density. Second, compared to PLZT 9/65/35 bulk ceramics, Maher [13] has shown that PLZT 12/70/30 ceramics have a substantially lower temperature coefficient of capacitance. Further, by using proper processing techniques we have been able to fabricate PLZT 12/70/30 films with dielectric constants in excess of 1000, which are comparable to those measured for near morphotropic phase boundary and PLZT x/65/35 films.

EXPERIMENTAL PROCEDURE:

PLZT 12/70/30 films were fabricated by a chemical solution deposition process termed the inverted mixing order process [14] derived from the work of Yi and Sayer. [15] The 0.4 M solutions were synthesized by blending zirconium-n-butoxide and titanium isopropoxide at ambient temperature, then adding acetic acid and methanol, which was followed by heating of the solutions to 80°C and the addition of lead acetate followed by lanthanum acetate addition. Additional acetic acid, methanol, and distilled water were added to the solution to control viscosity and improve solution stability. Individual solution layers were spin deposited at 3000 RPM for 30 seconds on sputter deposited 170 nm Pt//40 nm Ti coated Si wafers. Highly (111) oriented Pt bottom electrodes were obtained. Pyrolysis temperatures were varied from 300°C to 550°C with a soak time of 3 minutes being held constant throughout our study. Typically, four layers were deposited which resulted in a final PLZT film thickness of approximately 400 nm. Final crystallization was achieved by heating the films at 50°C/min to 650°C and holding for 30 minutes. Solution synthesis and film deposition were performed in laminar flow hoods.

The top electrodes used for our PLZT films were for the majority of cases 100 nm thick Pt films of 0.7 mm diameter. Select PLZT film coated wafers were deposited with 1mm, 3mm and/or 6 mm diameter top electrodes. Yields (not electrically shorted) ranged from 50% to 80% for the 6 mm diameter top electrodes. Capacitance and dissipation factor for our PLZT film samples were measured using an HP 4284A precision LCR meter at frequencies ranging from 100 Hz to 1 MHz. Polarization versus electric field data were measured using a Radiant Technologies Precision Workstation. Typically, dielectric hysteresis characteristics were measured at a frequency of 1 kHz.

Materials analysis consisted of SEM, TEM and X-ray diffraction analyses. TEM specimens in both cross-sectional and plan views were prepared using 5 kV argon ions. Ion milling at a 10° angle in a liquid nitrogen cold stage and ambient milling at an angle of 3.5° of the PLZT films were compared. No detectable film damage to be caused by the room temperature milling. Because of the lower milling angle, higher quality images over a greater thickness range could be obtained. Specifically, high quality views that included the entire PLZT film cross section and the bottom electrode were achieved by 3.5° milling; whereas, only 150 nm depths could be readily imaged using 10° milling angles.

Results and Discussion

The effects on PLZT film microstructure and dielectric properties were studied for the following process parameters: Pb stoichiometry, pyrolysis temperature, platinum vs. perovskite structure electrodes, seed layers, PbO atmosphere control, rapid thermal processing and PbO overcoat layers. Two of the most critical parameters for formation of single phase perovskite thin films: pyrolysis temperature and PbO overcoat layers, will be emphasized in this report. The effects of process parameters on dielectric constants of the PLZT films are shown in Figure 1. When standard 10 mol% excess Pb, 300°C pyrolysis temperature and no PbO overcoat layer processing was used, then a PLZT 12/70/30 film with a 1 kHz dielectric constant of only 250 was obtained (not shown in Figure). For comparison, for NMPB films, such as PZT 52/48, process procedures that include 10 mol% excess Pb addition and a 300 °C pyrolysis result in single phase perovskite films with dielectric constants on the order of 1000. By increasing the excess Pb content to 20 mol% a 1 kHz dielectric constant of 400 was achieved for the PLZT 12/70/30 films. Use of a $PbTiO_3$ seed layer increased the 1 kHz dielectric constant to 600, while use of a PbO overcoat layer with no seed layer increased the dielectric constant to 840. A process that included both a PbO overcoat layer and a 550°C pyrolysis temperature enhanced the 1 kHz dielectric constant to 1200 as shown in Figure 1.

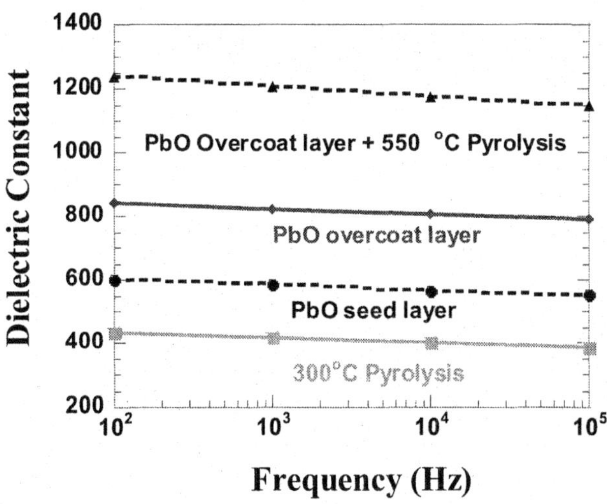

Figure 1. Dielectric constant versus frequency for 20 mol% excess Pb containing PLZT 12/70/30 films for different processing parameters.

Later studies indicated that dielectric constants of 1100 or greater could be achieved without the use of a PbO overcoat layer if pyrolysis temperatures of 550°C were used. The best films of our study that exhibited the highest dielectric constant and highest energy density were fabricated using 20% to 30% excess Pb additions, 550°C pyrolysis temperatures, and either with or with out PbO overlayers.

Single phase perovskite films were observed for all PLZT 12/70/30 films with dielectric constants in excess of 1000. Interestingly, the structures of the 550°C pyrolyzed films were very different using PbO top layer versus no PbO overlayer processing. A cross-sectional TEM micrograph of a 550°C pyrolyzed film with no PbO overlayer and 30% excess Pb addition is shown in Figure 2. While the PLZT film has separated from the underlying Pt layer during TEM preparation in Figure 2, it is clearly evident that the PZT film microstructure consists of fine, less than 0.2 μm wide, perovskite grains that extend from the bottom platinum electrode to the top surface of the film. There is no evidence of fine grained, low dielectric constant (K = 50) fluorite phase on the top film surface. If it does exist, it is less than 1 nm thick. The PLZT 12/70/30 film that is pyrolyzed at 550°C and has a PbO overcoat layer shown in Figure 3 is also single phase perovskite, but the structure is substantially different. While columnar perovskite grains emanating from the bottom electrode are evident, the top of the film consists of large diameter (1 μm) rosette perovskite structures. We postulate that the large perovskite grains nucleate from the top film surface, perhaps from PbO crystals that are derived from the PbO overlayer. The top surface

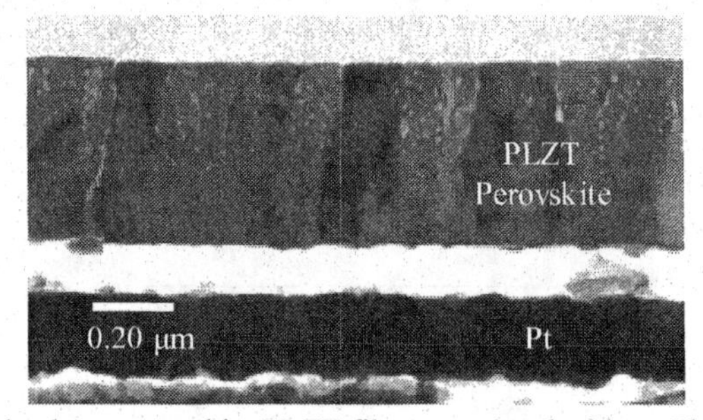

Figure 2: Single phase perovskite PLZT films processed with no PbO overlayer exhibiting columnar perovskite grains from the bottom electrode to the top of the PLZT film.

Figure 3: Single phase perovskite films processed with PbO top layer exhibiting rosette perovskite grains in top half of the film.

perovskite grains appear to grow down through the fluorite phase and impinge on the columnar perovskite grains growing up from the bottom electrode. Possible reasons for the larger top surface perovskite grains are that the top film surface has a lower nucleation density compared to the relatively rougher bottom Pt electrode, or that perovskite growth is enhanced in the Pb-rich environment of the top surface. Previous studies [16] indicated an enhanced perovskite nucleation rate due to enhanced nucleation surface dimensionality on rougher surfaces, which lowers the activation energy for perovskite phase nucleation. An interesting observation was that while both perovskite films exhibited preferential (111) orientation, the PLZT film with rosettes exhibited greater (111) orientation by X-ray diffraction.

HZPLZT films with lower dielectric constants, less than 1000, had evidence of second phase fluorite, and thus, were not single phase perovskite. Figure 4 shows that the volume percentage of the low dielectric constant fluorite phase is much greater for a PLZT 12/70/30 film with a dielectric constant (K) of 180 film compared to a film with a dielectric constant of 650. While both films were pyrolyzed at 300°C, the K = 180 film was processed with no excess Pb addition, while the K = 650 film solution had 30 mol% excess Pb addition. For the film with a measured dielectric constant of 650, there is evidence of an approximately 20 nm to 60 nm thick fluorite layer at the top of the PLZT film. The fluorite layer covers a majority of the film surface but there are isolated regions for which the perovskite phase reaches the top film surface. If one assumes a dielectric constant of 50 for the fluorite layer and a dielectric constant of 1200 for the major phase perovskite, then a dielectric constant of 560 is calculated using series dielectric mixing rules and assuming a uniform 20 nm thick fluorite layer. For the K = 180 film, the fluorite volume fraction is far more extensive than for the K = 650 film. In many instances, the fluorite phase extends from the top to the bottom of the PLZT film.

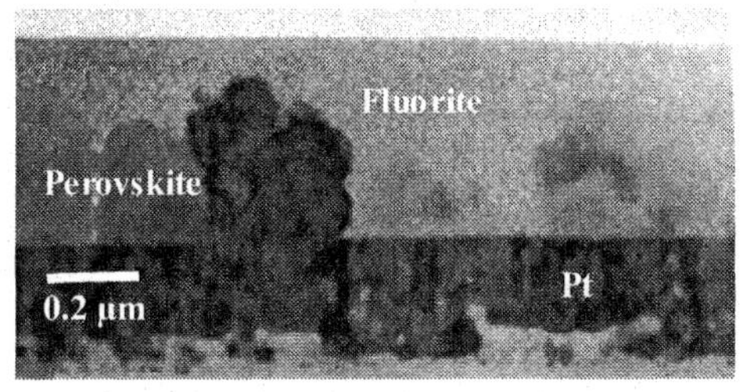

(a)

(b)

Figure 4: Low dielectric constant (a) K = 180 PLZT film, bright field TEM image and (b) K =650 PLZT film dark field TEM image.

Dielectric hysteresis characteristics of three different single phase perovskite films measured with 600 kV/cm, 1 kHz applied fields are shown in Figure 5. The three compositions tested were PLZT 5/50/50, PNZT 5/50/50 and PLZT 12/70/30. The PNZT 5/50/50 film denotes a 5 mol% Nb addition. The PLZT 5/50/50 and PNZT 5/50/50 films were fabricated several years ago as part of a National Laboratory – Industry consortium. While both of the near morphotropic phase boundary films saturate at lower fields, the PLZT 12/70/30 film has much more linear dielectric behavior with applied field and also has a higher polarization of 56 $\mu C/cm^2$ than the other two films. Polarizations of 32 and 42 $\mu C/cm^2$ at 600 kV/cm were measured for the PLZT 5/50/50 and PNZT 5/50/50 films, respectively. Energy densities (E_D) of these films were calculated using two different methods, for which the difference in calculated E_D values was less than 3 percent. The first method used linear dielectric constant approximations for four different segments of the first quadrant polarization versus field curves. The integral quantity $E_D = \int PdE = \frac{1}{2}\, \varepsilon\, E^2$ was used to calculate the energy density corresponding to each segment and then the four different segment energy densities were added to determine the total energy density. For the second method, the first quadrant electric field versus polarization characteristic was integrated, since the stored energy for a nonlinear capacitor [17] is equal to the area under the voltage versus charge curve. Energy density values obtained were 6.4 J/cm^3 for the PNZT 5/50/50 film, 5.4 J/cm^3 for the PLZT 5/50/50 film and 11.9 J/cm^3 for the PLZT 12/70/30 film. These energy density values are for the PLZT dielectric only and do not include electrode and substrate volumes.

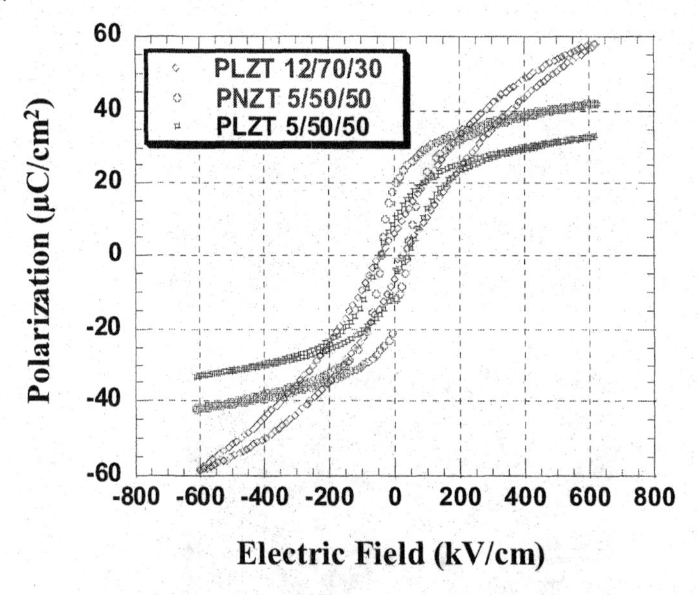

Figure 5: Dielectric hysteresis characteristics of PLZT 5/50/50, PNZT 5/50/50and PLZT 12/70/30 films.

Dielectric hysteresis characteristics were measured and energy densities were calculated for fields that were roughly 90% of the breakdown field level for PLZT 5/50/50, PNZT 5/50/50 and PLZT 12/70/30 films. These higher field measurements provide a metric for the three dielectric films for short term DC or pulse applications. For electrical breakdown strength measurements all three samples were of similar geometry and were tested under identical electrical conditions. These films were of 0.4 to 0.6 μm thickness and had electrodes of 0.7 to 1 mm diameter. A 250 μsec unipolar ramp voltage was used to obtain breakdown strengths for all three films. The breakdown strengths were 0.8 MV/cm, 1.1 MV/cm and 1.5 MV/cm, for the PNZT 5/50/50, PLZT 5/50/50 and PLZT 12/70/30 films, respectively. A dielectric hysteresis characteristic is shown in Figure 6 for a single phase perovskite PLZT 12/70/30 film for a 1.4 MV/cm applied field. The applied field is just below the breakdown strength field and is more than twice the value of the applied field for the films in Figure 5. The calculated energy density for the PLZT 12/70/30 film for the 1.4 MV/cm field was 29 J/cm^3. The increase in energy density compared to that for the 600 kV/cm hysteresis characteristic is a result of the quadratic field dependence of stored energy that overshadows the saturation of the dielectric constant with field. An effective dielectric constant of 250 was calculated for the linear polarization versus field approximation over the 0.85 MV/cm to 1.4 MV/cm electric field range. It must be noted that larger diameter (6 mm) electrodes reduced the polarization measured at 1.4 MV/cm from 60.5 $\mu C/cm^2$ to 50 $\mu C/cm^2$, presumably due to the reduction of fringe field effects. A 30% reduction in energy density from 29 J/cm^3 to 20 J/cm^3 results from the lower polarization values. Inclusion of the top and bottom Pt electrodes as part of the volume of the original 0.7 mm diameter capacitor results in an energy density of 15 J/cm^3 for the 1.4 MV/cm field. This energy density value is among the highest reported energy densities to date for a PZT based thin film capacitor. For comparison, the PNZT 5/50/50 and PLZT 5/50/50 films had energy densities, for the dielectric only, of 14 and 7 J/cm^3, respectively, for fields that were roughly 90% of the breakdown value.

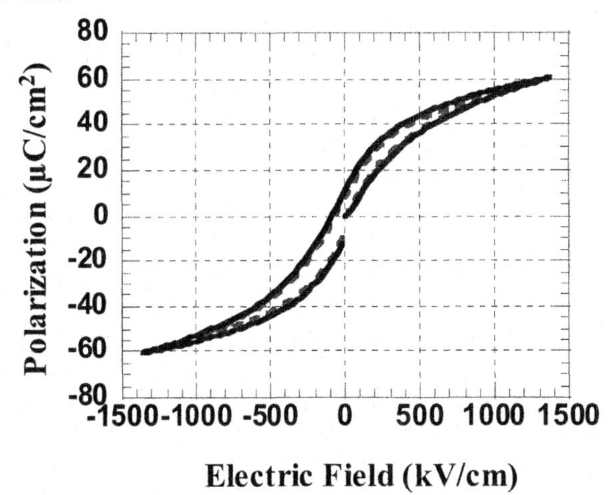

Figure 6: Polarization versus electric field hysteresis characteristics for 1.0MV/cm and 1.4 MV/cm applied electric fields for PLZT 12/70/30 thin film.

SUMMARY

High Zr content PLZT 12/70/30 films were fabricated that exhibited greater linearity and higher energy densities at high fields than La and Nb doped near morphotropic phase boundary films. However, processing requirements to achieve single phase perovskite thin films were found to be more stringent than for near morphotropic phase boundary films. More excess Pb, higher pyrolysis temperatures and use of PbO overcoat layers all proved beneficial in the development of single phase perovskite PLZT 12/70/30 films. Single phase perovskite PLZT 12/70/30 films that were fabricated with and without a PbO overlayer had substantially different grain structures. While single phase perovskite films with no overlayer were fine grained (roughly 0.15 micrometer dimension) and columnar from the bottom electrode to the top of the film, films fabricated with a PbO over layer exhibited large micrometer dimension perovskite crystals that comprised the top half of the film. Our work indicates that high Zr content PLZT thin film dielectrics show great promise for enhancing the volumetric efficiency of capacitors for integrated microsystems.

ACKNOWLEDGMENTS

Jill Wheeler, Walter Olson and Chuck Hills are acknowledged for outstanding technical assistance. Sandia is a multiprogram laboratory operated by Sandia Corporation, a Lockheed Martin Company, for the United States Department of Energy under contract DE-ACO4-94AL85000.

REFERENCES

[1] D. Dimos, S. J. Lockwood, R. W. Schwartz and M. Steven Rodgers, "Thin-Film Decoupling Capacitors for Multichip Modules, *IEEE Trans on Components Packaging and Manufacturing Technology*, Part A. **18** [1] 174- 79 (1994).

[2]D. Dimos, S.J. Lockwood, T.J. Garino, H.N. Al-Shareef and R.W. Schwartz, "Integrated Decoupling Capacitors Using $Pb(Zr,Ti)O_3$ Thin Films, Materials Research Society Symposium, **433** 305–16, (1996).

[3]B. A. Tuttle, D.P.Williams, T. Headley, M. Rodriguez, J. Wheeler, J. Voigt, and G. Brennecka, "High Zr Content PLZT Films for MEMS Pulse Discharge Applications," 15th Annual Rio Grande Symposium, Albuquerque, NM October 15, 2003.

[4]J. P. Maria, K. Cheek, S.K. Streiffer, S-H. Kim,G. Dunn and A. Kingon, "Lead Zirconate Titanate Thin Films on Base Metal Foils: An Approach for Embedded High-Permittivity Passive Components," *Journal of the American Ceramic Society*, **84** [10] 2436-38, 2001.

[5]F. Xu, S. Trolier-McKinstry, W. Ren, B. Xu, Z-L. Xie, K.J. Hemker, "Domain Wall Motion and its Contribution to the Dielectric and Piezoelectric Properties of Lead Zirconate Titanate Films," *Journal of Applied Physics*, **89** [2] 1336-48 (2001).

[6]D. Liu and J. Mievissen, 'Thick Layer Deposition of Lead Perovskites Using Diol-Based Chemical Solution Approach," *Integrated Ferroelectrics*, **18**, 263-74 (1997).

[7]D-J. Kim, D. Y. Kaufman, S. K. Streiffer, T.H. Lee, R. A. Erck and O. Auciello, "Chemical Solution Deposition of PLZT Films on Base Metal Foils," Ferroelectric Thin Films XI, Materials Research Society Proceedings, **748** 457-62 (2003).

[8]B. Tuttle, T. Headley, B. Bunker, R. Schwartz, T. Zender, C. Hernandez, D. Goodnow, R.Tissot and A. Carim, "Microstructural Evolution of $Pb(Zr,Ti)O_3$ Thin Films

Prepared by Hybrid Metallo-organic Decomposition," *Journal of Materials Research*, **7** [7] 1876 – 82 (1992).

[9]B.A. Tuttle and R.W. Schwartz, "Solution Deposition of Ferroelectric Thin Films," *Materials Research Society Bulletin*, **21** [6] 49-54 (1996).

[10]C.D.E. Lakeman, Z. Xu, and D. Payne, "On the Evolution of Structure and Composition in Sol-Gel-Derived Lead Zirconate Titanate Layers," *J. Mater. Res.*, **10** [8] 2042-51 (1995).

[11]R. Holman and R. Fulrath, "Intrinsic Nonstoichiometry in the Lead Zirconate-Lead Titanate System Determined by Knudsen Effusion," *J.Appl.Phys.*, **44** [12] 5227-36 (1973).

[12]M. Mandeljc, B. Malic, M. Kosec and J. Dravic, "Crystallization of Zirconium-Rich PLZT Thin Films Below 500°C," Abstracts for the International Joint Conference on the Applications of Ferroelectrics, Nara, Japan 154 (2002)

[13]G. Maher, "A New PLZT Dielectric with Characteristics of X7R Multilayer Capacitors," Proc. 33rd Electronics Components Conference, IEEE Orlando, FL, 173-77 (1983).

[14]R. Schwartz, B. Bunker, D. Dimos, R. Assink, B. Tuttle, D. Tallant, I. Weinstock, and D. Haaland, "Solution Chemistry Effects in $Pb(Zr,Ti)O_3$ Thin Films," Proc. Of the 3rd Int. Symposium on Int. Ferroelectrics, 535-46 (1991).

[15]G. Yi, Z. Wu, and M. Sayer, "Preparation of $Pb(Zr,Ti)O_3$ Thin Films by Sol-Gel Processing: Electrical, Optical and Electro-optic Properties," *Journal of Applied Physics*, **64** [5] 2717-24 (1988).

[16]B. A. Tuttle, "$Pb(Zr,Ti)O_3$ Based Thin Film Ferroelectric Nonvolatile Memories"; pp. 245-66 in *Thin Film Ferroelectric Materials and Devices* edited by R. Ramesh, Kluwer Academic Publishers, Boston,1997.

[17]P.M. Chirlian, pp. 47-48 in *Basic Network Theory*, McGraw-Hill Book Company, New York, 1969.

RELIABILITY STUDY ON SPUTTER DEPOSITED BARIUM STRONTIUM TITANATE THIN FILM CAPACITORS

Nobuo Kamehara, J. D. Baniecki, T. Shioga and K. Kurihara
Fujitsu Laboratories Ltd.
10-1 Morinosato-Wakamiya, Atsugi 243-0197, Japan

ABSTRACT

Barium Strontium Titanate (BST) is one of the best materials for thin film capacitor applications. Many studies on electric properties of BST thin films were reported, however, understanding of reliability and degradation mechanism of BST thin film capacitor is not enough. In this study, fundamental electrical and reliability properties of BST thin film capacitors were investigated. 200 nm thick BST thin films were deposited on Pt/Si substrates at 500°C by RF magnetron sputtering. After top Pt electrodes were deposited and patterned, samples were annealed at 500°C under O_2 atmosphere. Basic electrical properties were a capacitance density of 1.8 $\mu F/cm^2$, leakage current density < 10^{-9} A/cm^2 at 1.5V, and a breakdown voltage > 30V at 20°C. The projected mean time to failure for 1.5 volt operation is extrapolated to be in excess of 10^4 years at 75°C and 16.2 years at 125°C. The physical mechanisms contributing to capacitor failure are interpreted to be due to ionic migration and charge injection.

INTRODUCTION

Barium strontium titanate (BST) is one of the best materials for thin film capacitors, because of its high dielectric permittivity, low dielectric loss, and good leakage properties. Because of these excellent properties, extensive studies of BST thin films have been done for potential applications to DRAM, decoupling capacitors, MMIC and MCM [1]-[5]. Deposition processes and electrical properties of BST thin films have also been widely reported using MOD, CVD, sputtering, PLD and so on. However, understanding of reliability and degradation mechanism of BST thin film capacitor is not enough. In this study, fundamental electrical and reliability properties of sputter deposited BST thin film capacitors were investigated.

EXPERIMENTAL

BST thin films of thickness about 200 nm were deposited on $Pt/TiO_2/SiO_2/Si$ substrates by using an RF magnetron sputtering technique from a $(Ba_{0.7}Sr_{0.3})TiO_3$ target at substrate temperatures of 500°C. Crystallinity of the deposited films were characterized using X-ray diffraction. Microstructure and thickness of the films were determined by TEM and SEM. For capacitance measurements, Pt top electrodes were sputter deposited, and patterned by using Ar ion milling. Post annealing was then done in flowing O_2 at temperatures below the BST deposition temperature. Capacitances were measured from 100Hz to 40MHz by an HP4194A impedance gain phase analyzer with A. C. oscillation level of 50 mV. The leakage currents were recorded using an HP4339B high resistance meter using a voltage step technique that consisted of stepping the voltage in equally spaced voltage increments up to the maximum value. A hold time of 100 seconds was used between each voltage step after which the leakage was recorded. Time dependent dielectric breakdown (TDDB) and leakage currents were recorded using an Kiethley 2400 source meter.

RESULTS AND DISCUSSION

BST Microstructure

Figure 1 shows a cross sectional TEM image of a Pt/BST/Pt thin film capacitor. The BST is 200 nm thick and was deposited by RF Magnetron sputtering at a deposition temperature of 500°C. The BST film has a columnar microstructure and smooth surface. No interfacial layers at top and bottom electrodes and also no grain boundary layers were observed. X-ray diffraction showed (100) preferential orientation. The reason of very flat surface in spite of columnar micro structure seems to the (100) preferred orientation.

Electrical Properties

Figure 2 shows the C-V characteristic for a Pt/BST/Pt thin film capacitor. The capacitance density is 1.8 $\mu F/cm^2$ at zero D.C. bias, and the bias dependence was typical for polycrystalline BST thin films.

Figure 3 shows temperature dependent the J-V characteristics for the same capacitor as shown in Figure 2. The capacitor exhibits very good leakage properties achieving a current density of 10^{-9} A/cm^2 at 2 volts and 103°C. The breakdown voltage at room temperature, measured using a stair case J-V measurement with a 100 s delay time, was over 30V.

Figure 4 shows the temperature dependent Schottky plots. Linear relationships shown in the figure indicate the leakage mechanism of 500°C deposited BST thin film follows an interface-controlled Schottky type mechanism [6] that is usually observed in polycrystalline BST thin films.

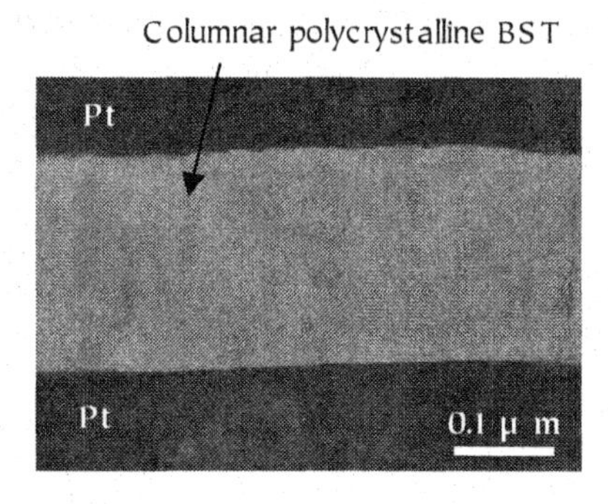

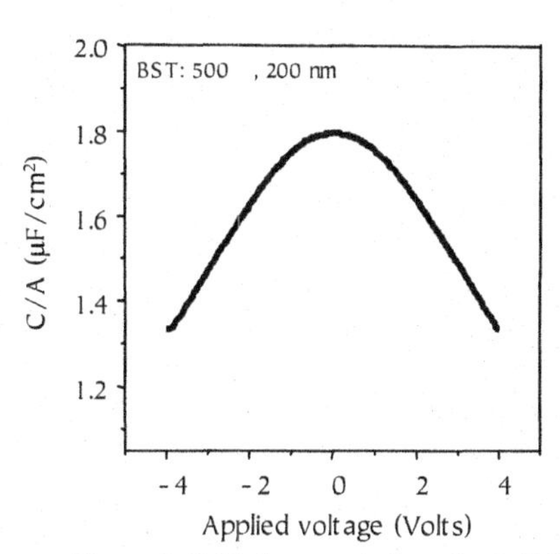

Figure 1 Cross-sectional TEM image of 500 deposited 200 nm thick BST thin film capacitor.

Figure 2 C-V characteristics of a Pt/BST/ Pt thin film capacitor.

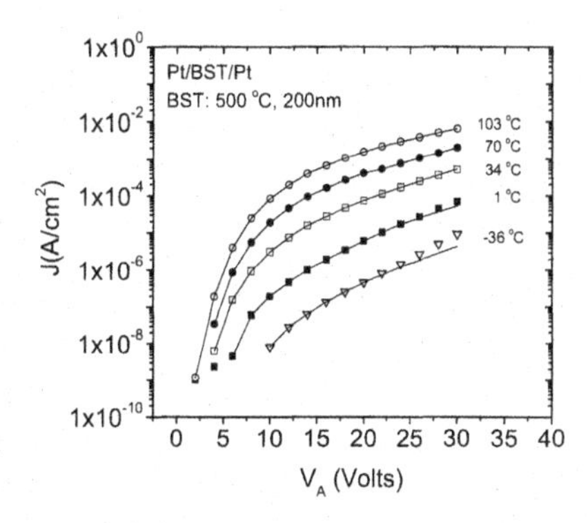

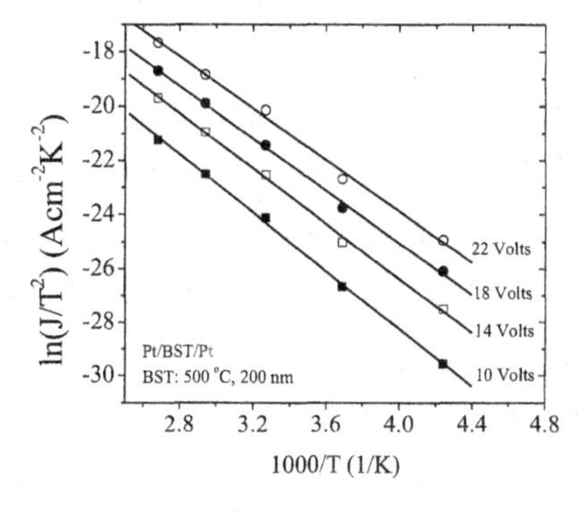

Figure 3 J-V characteristics of a Pt/BST/Pt thin film capacitor.

Figure 4 Schottky plot (ln[J/T2] versus 1000/T) of the data set shown in Fig. 3.

BST capacitor reliability

For the basic reliability study, time dependent dielectric breakdown (TDDB) measurements [7] were performed. The life time of the capacitors were defined that failure occurs when the leakage current increases by a factor of 10 from its minimum value.

Figure 5 shows the current time characteristics of Pt/BST/Pt thin film capacitors measured at a temperature of 379 °C for applied step voltages of 1.5, 5, and 10 Volts. As shown in Fig. 5 the current density at breakdown (catastrophic capacitor failure) and the distribution in characteristic time to failures depend strongly on the applied field.

The distribution of time to failures at different temperatures and voltages were analyzed using a Weibull distribution and the mean time to failure (MTTF) subjected to a 1.5 V were calculated. Figure 6 shows the MTTF versus 1000/T for Pt/BST/Pt thin film capacitors. The projected MTTF for 1.5 volt operation is 16.2 years at 125°C showing good reliability for BST thin film capacitors deposited by sputtering.

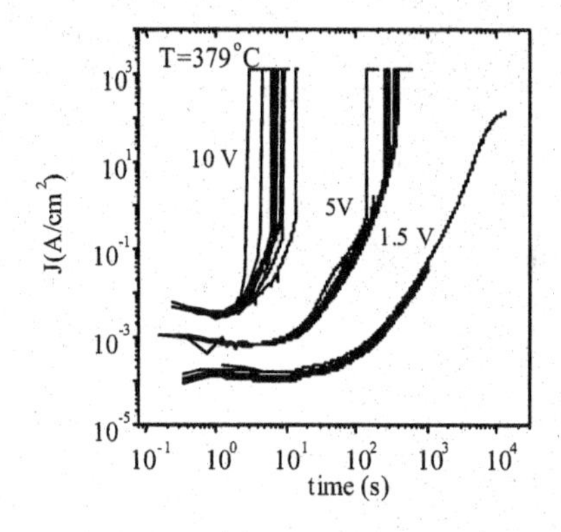

Fig. 5. Current time characteristics of Pt/BST/Pt thin film capacitors.

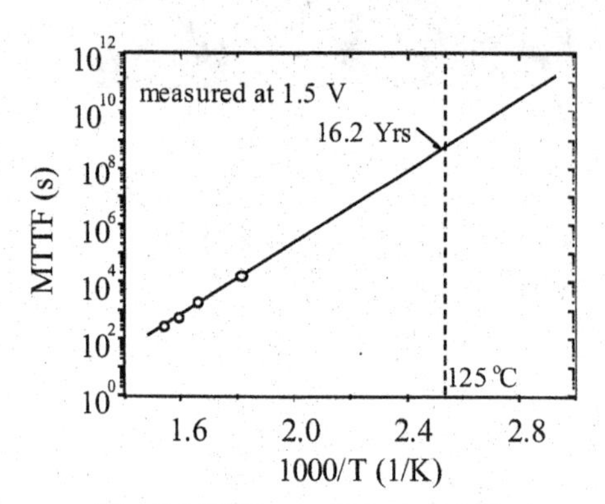

Fig. 6. MTTF vs. 1/T for Pt/BST/Pt thin film capacitors.

As shown in Fig. 6 the MTTF follows an Arrhenius-type temperature activation with the activation energy E_A=1.29 eV. E_A is similar to the enthalpy of migration for oxygen vacancies in $SrTiO_3$ thin films [8] suggesting that the degradation in leakage properties may be mediated by the drift of positively charged oxygen vacancies to the cathode under the action of the applied bias.

CONCLUSION

Fundamental electrical and reliability properties of BST thin film capacitors were investigated. 200nm thick BST thin films deposited by RF magnetron sputtering at 500°C achieved C/A > 1 μF/cm^2 and leakage current density < 10^{-9} A/cm^2 at 100°C. The projected mean time to failure for 1.5 volt operation is extrapolated to be in excess of 10 years at 100°C. This study demonstrates that BST thin film capacitors are very reliable even at high temperature operation and useful for the DRAM and decoupling applications.

REFERENCES

[1]T. S. Kim and C. H. Kim, "Structural and electrical properties of rf magnetron sputtered $Ba_{1-x}Sr_xTiO_3$ thin films on indium-tin-oxide coated glass substrates", J. Appl. Phys., vol.75, pp. 7998-8003, 1994

[2]H. Kobayashi and T. Kobayashi, "Heteroepitaxial growth of quaternary $Ba_{1-x}Sr_xTiO_3$ thin films by ArF excimer laser ablation", *Jap. J. Appl. Phys.*, vol. 33, pp. L533-L536, 1994

[3]D. Tahan, A. Safari, L. C. Klein, "Sol-gel preparation Barium-Strontium Titanate thin film", *International Symposium on Application of Ferroelectrics*, 1995, pp. 427-430

[4]J. D. Baniecki, R. B. Laibowitz, T. M. Shaw, C. Parks, J. Lian, H. Xu Q. Y. Ma, "Hydrogen induced tunnel emission in Pt/$Ba_{1-x}Sr_xTiO_3$/Pt thin film capacitors, *J. Appl. Phys.* vol. 89, pp. 2873-2885, 2001.

[5]K. Kurihara, T. Shioga, J. D. Baniecki, "Electrical properties of low-inductance barium strontium titanate thin film decoupling", *Journal of the European Ceramic Society*, **24** 1873-1876 (2004).

[6]J.D. Baniecki, T. Shioga, K. Kurihara, N. Kamehara, "Investigation of the importance of interface and bulk limited transport mechanisms on the leakage current of high dielectric constant thin film capacitors", *J. Appl. Phys.* **94** [10] 6741(2003).

[7]S. Saha and S.B. Krupanidhi, "Microstructure related influence on the electrical properties of pulsed laser ablated (Ba, Sr)TiO_3 thin films", *J. Appl. Phys*, **88**[6] 3056-3513 (2000).

[8] R. Waser, J. Am. Ceram. Soc. 74, 1934 (1991)

RAMAN STUDY OF EFFECTS OF EXCESS BI CONTENT IN METALORGANIC DERIVED $Bi_4Ti_3O_{12}$ FILMS

C.Y. Yau, R. Palan, K. Tran and R.C. Buchanan*
Department of Chemical and Materials Engineering, University of Cincinnati, Cincinnati OH 45221-0012, USA

ABSTRACT

$Bi_4Ti_3O_{12}$ (BIT) films with different excess Bi content (0 – 6 mol.%) are studied using x-ray diffraction (XRD), scanning electron microscopy (SEM) and Raman scattering. From SEM data, the grain size and densification increases with excess Bi content, indicating that the grain growth of the film is related to the excess Bi content and liquid phase sintering occurs at 6 mol.% excess Bi BIT films. From the XRD data, the *c*-orientation of the BIT films increases from 0-4 mol.% and is well developed when Bi excess reaches between 4 and 7 mol.%. From the analysis of the Raman spectra, the Bi_2O_2 layer remains unchanged in the Bi-deficient (~ 0-2 mol.% Bi excess) as well as stoichiometric (~ 4-6 mol.% Bi excess) films; implying that during film growth, Bi first goes to the Bi_2O_2 layer in which the Bi site has higher stability than that in the perovskite-like slab. With increasing excess Bi, the excess Bi is incorporated into the A site in the perovskite-like slab, enhancing the distortion of TiO_6 octahedra. Also the preferred film orientation becomes *c*-direction. A decrease of the distortion of TiO_6 octahedra (e.g. by doping with a smaller ion like La) is possible and is important for improving the polarization in the *c*-direction of BIT films.

INTRODUCTION

Ferroelectric $Bi_4Ti_3O_{12}$ (BIT)[1,2] has low coercive field (~ 3.3 kV/cm) and dielectric constant along the *c*-axis, making *c*-oriented BIT a superior candidate for applications like low power consumption, high speed electronic memory devices. The crystal structure of BIT consists of a $(Bi_2Ti_3O_{10})^{2-}$ slab composed of three perovskite-like units of nominal composition $BiTiO_x$, being sandwiched between two $(Bi_2O_2)^{2+}$ layers along the *c*-direction, as shown in Fig. 1. The Bi content is known to be important to the improvement of polarization-related properties such as orientation and microstructure, etc. of BIT film.[3,4,5] However, the issues of which site the excess Bi is incorporated and how excess Bi affects the orientation and polarization of BIT films are still unresolved. To resolve these questions, we investigated the effect of excess Bi on the crystal orientation and lattice dynamics of BIT films by x-ray diffraction (XRD), scanning electron microscopy (SEM) and Raman scattering.

EXPERIMENTAL

In the present study, BIT films with 0-10 mol.% Bi excess were fabricated under the same conditions on (111)Pt/Ti/SiO_2/Si substrates by metalorganic decomposition (MOD) method.[5] The phase development and orientation of the films were studied using XRD θ-2θ scan. The surface morphology of samples was studied using SEM. Raman scattering in samples with 0-6 mol.% Bi excess (other films are not studied using Raman scattering because the film contains a second

* Author to whom correspondence should be addressed; phone: 513-556-3190; fax: 513-556-1539; electronic mail: relva.buchanan@uc.edu

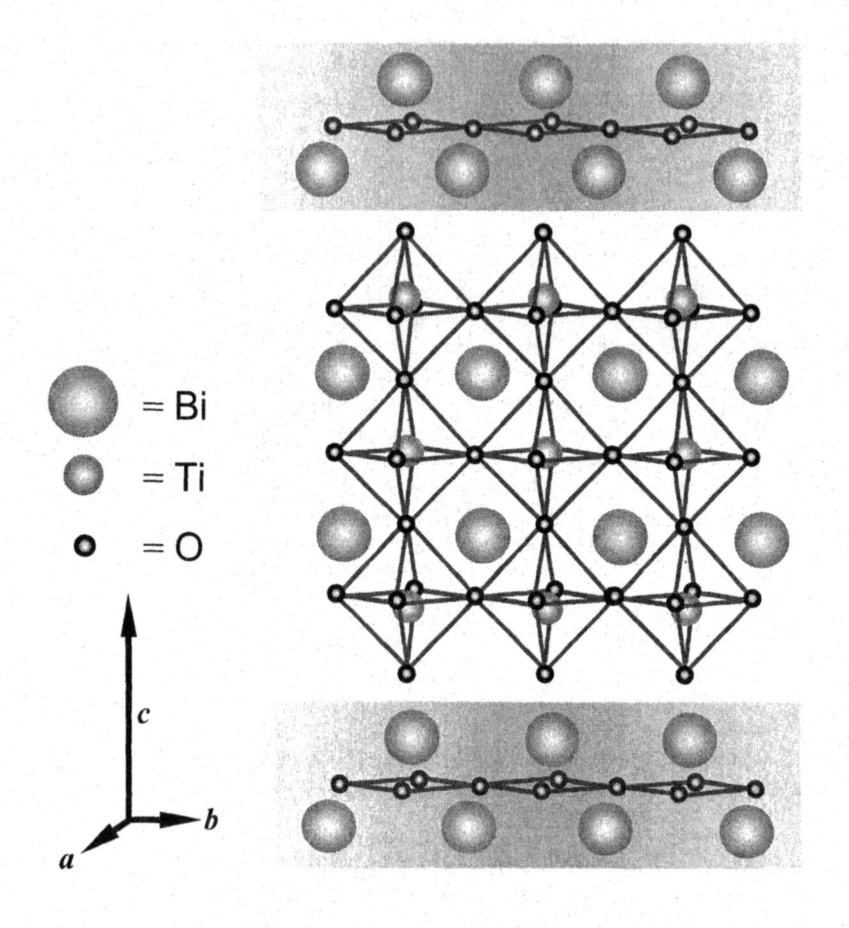

Fig. 1. Undistorted structure of $Bi_4Ti_3O_{12}$. The shaded parts are Bi_2O_2 layers which play an important role in rigid layer vibration modes (after Graves *et al.*[6]).

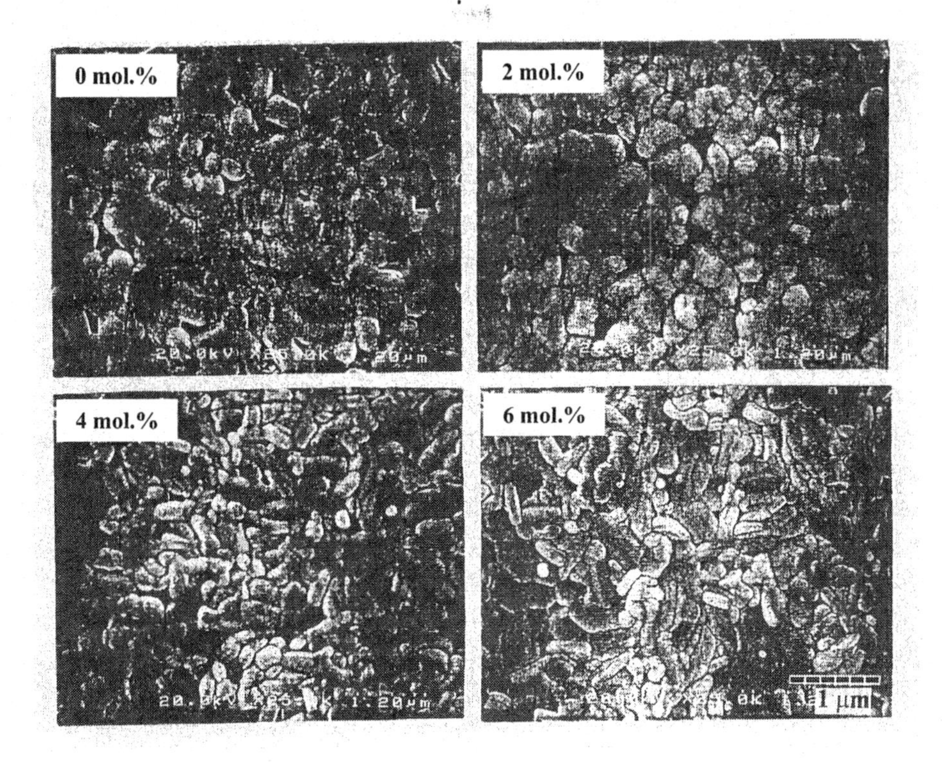

Fig. 2. SEM surface images of $Bi_4Ti_3O_{12}$ films with different mol.% Bi excess.

phase as Bi excess is larger than 6 mol.%, making the mode analysis confused) was performed at room temperature using T64000 triple monochrometer system from instruments SA equipped with a charge-coupled device (CCD) detector and a microscope attachment; and the 647.1 nm line Kr^+ ion laser was used to excite the scattering.

RESULTS AND DISCUSSIONS

From SEM pictures, as shown in Fig. 2, the grain size increases with excess Bi content. Necking between particles increases significantly from 0 to 2 mol.% excess Bi content. Further, the grains become slender plates in 4 – 6 mol.% excess Bi BIT films. This indicates that the grain growth in the excess Bi BIT film is related to the excess Bi content. The pictures indicate that liquid phase sintering may occur in 6 mol.% excess Bi BIT films.

Fig. 4 shows the Raman spectra and Fig. 5 shows the Raman mode frequency shift of BIT films with different mol.% Bi excess. For the paraelectric tetragonal symmetry of BIT above T_C, 16 Raman active modes ($6A_{1g} + 2B_{1g} + 8E_g$) can be expected;[6] for ferroelectric monoclinic symmetry below T_C, more phonon modes are expected,[7] e.g. in single crystal.[8] However, in Fig. 4

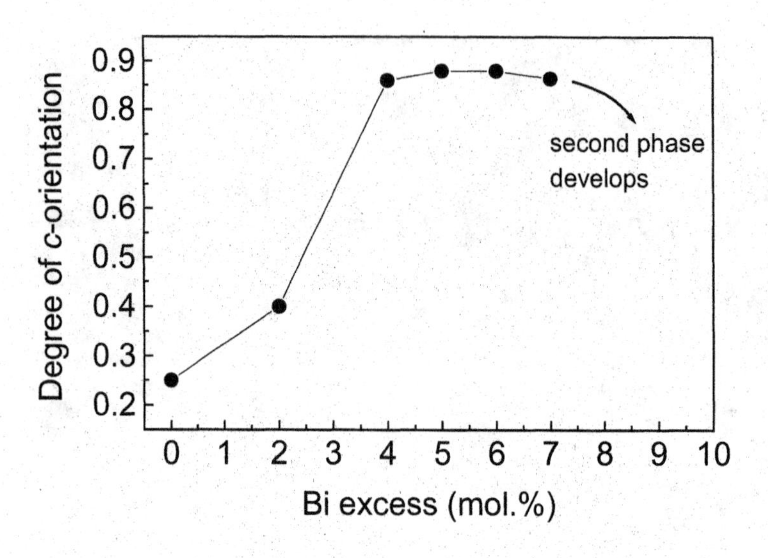

Fig. 3. Degree of *c*-orientation of BIT films with different mol.% Bi excess.

and 5 the number of modes of all films cannot fulfil the selection rule and this may be due to broken symmetry, mode-overdamping and mode-overlapping making some modes unidentified.

Comparing the Raman modes in 2-6 mol.% Bi excess films, the ones with 0 mol.% Bi excess are very different. First, the number of modes of the film with 0 mol.% Bi excess is less than that with Bi excess. Second, most modes of the 0 mol.% Bi excess film are overdamped and highly shifted while the dampings of those modes of films with Bi excess are decreased. These two features can be explained by our SEM study as follows. In Fig. 2, the grain size of 0 mol.% Bi excess films are ~ 0.1-0.2 μm which is less than the grain size of 0.36 μm for complete stress relaxation.[10] The splitting of the mode at ~ 550 cm^{-1} in 0 mol.% Bi excess film into two modes in 2 mol.% Bi excess film suggests a symmetry breaking effect. The possible stress induced symmetry breaking effect inside the grain could suppress the emergence of some Raman modes that belong to the BIT monoclinic symmetry; causing the number of modes in the film with 0 mol.% Bi excess lower than other films. The emergence of the unobserved modes in non-zero mol.% Bi excess films can be attributed to the stress relaxation in the 2-6 mol.% Bi excess films with large enough grain size; together with the decrease of structural disorder in the incorporation of Bi excess into BIT, the symmetry of BIT can be restored. The larger damping of the Raman modes in the film with 0 mol.% Bi excess than that in 2-6 mol.% Bi excess films is then due to the structural disorder caused by Bi deficient sites in BIT. The unobserved modes in 0 mol.% Bi excess film emerge in 2-6 mol.% Bi excess films making the overall features of the Raman spectra of such films (i.e. number and shape of the modes) similar to that of the BIT single crystal

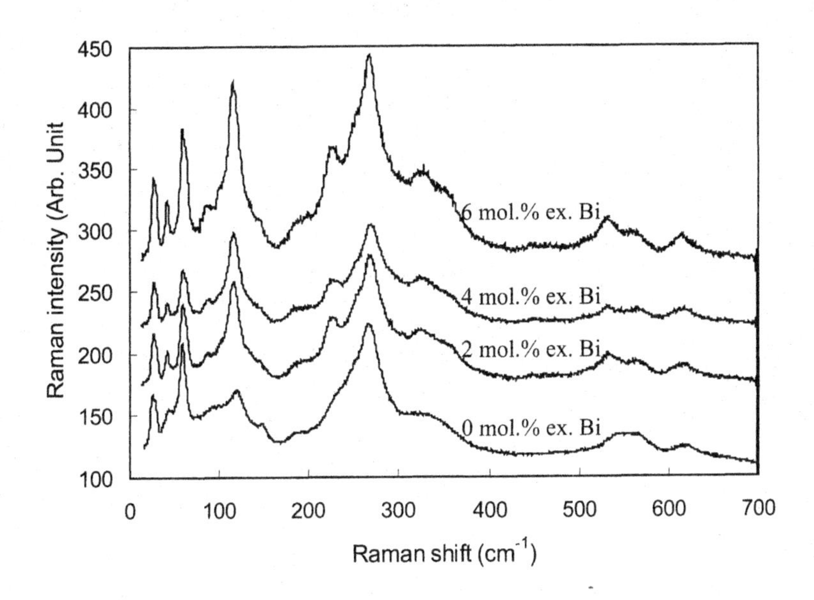

Fig. 4. Raman spectra of Bi excess (0-6 mol.%) BIT films.

reported by Kojima *et al.*[11] and this confirms the XRD peaks[9] that the BIT films are phase pure until below 7 mol.% Bi excess, at which second phase emerges.

The lowest frequency mode is observed at 25.2 cm^{-1} in 0 mol.% Bi excess film and at 26.46 cm^{-1} in 2, 4 and 6 mol.% Bi excess films respectively. This mode can be assigned as the soft mode as it softens and disappears in the phase transition of BIT.[8,11]. Comparing this mode with that in single crystal,[11] this mode in thin film is usually lower than that of single crystal because of the existence of residue stress in thin films. Since pressure-induced phase transition can cause a softening of this mode,[8] the upshift of this mode from 0-2 mol.% Bi excess films thus possibly indicates a stress relaxation.

In layered compounds like graphite[12] and Bi-layered compounds[13] there is a $\boldsymbol{q} \approx 0$ optic mode in which a layer moves very much like a rigid layer so that the averaged intralayer bond lengths are nearly unchanged and is called a rigid-layer mode.[12,14] The appearance of this mode does not depend on the point group symmetry of the crystal but is a consequence of the weakness of the interlayer bonding[12] and thus this mode can be observed in both ferroelectric and paraelectric phases. There are three such modes: one is a compressional mode (A_{1g} mode), also called a "breathing" mode, which is like a longitudinal wave propagating along the *c*-direction and is important in the superconductivity of layered perovskites;[15] the other one is a "ripple" mode (B_g mode) and it is like a transverse wave propagating parallel to the layer-plane direction; the third one is a "shear" mode (E_g mode) which is like a transverse wave (caused by the motions of layers

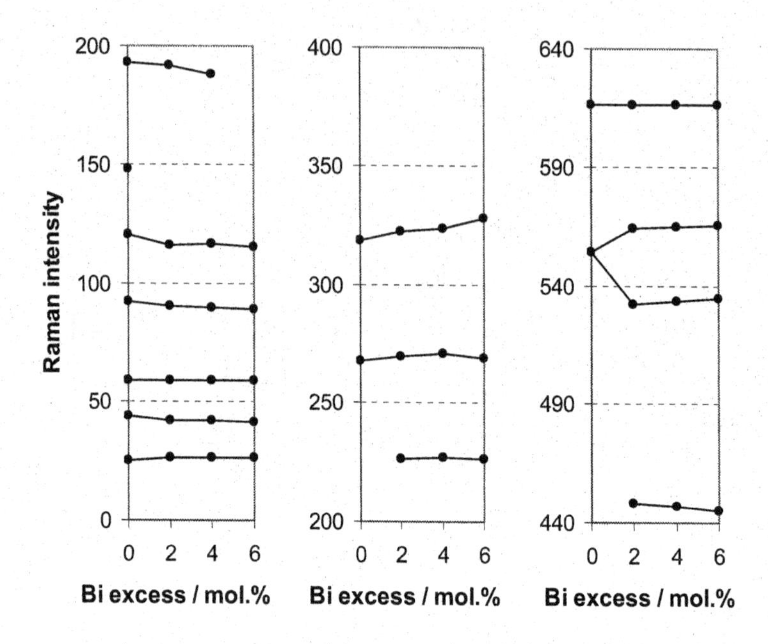

Fig. 5. Frequency shift of Raman modes of Bi excess (0-6 mol.%) BIT films. The lines are guides to the eye.

in the x(y) planes) propagating along the *c*-direction. The schematic representation of the three rigid layer modes is shown in Fig. 6. E_g mode has the lowest frequency and sometimes overlaps with other lowest frequency modes like the soft mode and the central mode; the "breathing" mode has the highest frequency. In BIT, the arrangement of the layers along *c*-direction is that –(Bi_2O_2)–($Bi_2Ti_3O_{10}$)–(Bi_2O_2)–($Bi_2Ti_3O_{10}$)–(Bi_2O_2)– where "–" represents a binding force which is much lower than the intralayer ionic bonds and "( )" represents a layer structure. Considering BIT having a E_g mode, the ($Bi_2Ti_3O_{10}$) and (Bi_2O_2) layers vibrate in a shear fashion against each other. Considering BIT having an A_{1g} "breathing" mode, the ($Bi_2Ti_3O_{10}$) and (Bi_2O_2) layers vibrate like a "breathing" motion in *c*-direction against each other. For these two modes, the change of reduced mass in either layer will have significant effect to the mode frequency. For "ripple" mode in BIT, the *c*-direction vibrations of atoms of a layer have antisymmetry for a 90° rotation about *c*-direction. Since the reduced mass of the ($Bi_2Ti_3O_{10}$) layer is much heavier than that of the (Bi_2O_2) layer, the change of reduced mass by incorporation of Bi in the ($Bi_2Ti_3O_{10}$) layer has little effect to this mode; but the change of reduced mass of (Bi_2O_2) layer by incorporation of Bi into this layer obviously has significant effect to this mode. In Fig. 4 and 5, the mode in 0 mol.% Bi excess film at 59.22 cm^{-1} which persists above T_C[11] is susceptible to be a rigid layer vibration mode, probably a "ripple" mode. A mode at 60 cm^{-1} is commonly observed to persist into the paraelectric phase in Bi-layered compounds and is commonly assigned as a rigid layer mode.[11,13] Rigid layer mode, independent of the crystal symmetry, stays at a fixed frequency below and

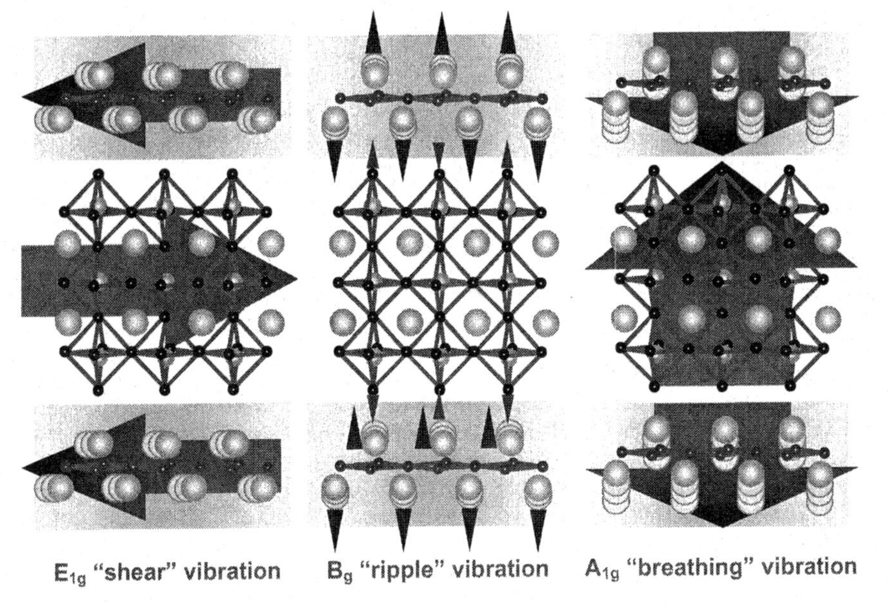

Fig. 6. Schematic presentation of the rigid layer modes of BIT. The arrows indicate the motion direction of the corresponding layer; the small arrows in the "ripple" vibration mode indicate the vibration directions of the corresponding atoms.

above T_C while other external modes below 100 cm^{-1} will either increase or decrease in frequency. Thus, the mode at 59.22 cm^{-1} can be assigned as a rigid layer mode. As we have discussed, the change of reduced mass in the perovskite-like slab have little effect in the "ripple" mode while the change of reduced mass in the Bi_2O_2 layer has significant effect in all three modes. Then the unchanged 59.22 cm^{-1} "ripple" mode implies that there is no change in the Bi_2O_2 layer. Then the Bi_2O_2 layer is intact in all the Bi deficient and Bi saturated films and this implies that Bi has a high preference to go into the Bi_2O_2 layer rather than the perovskite-like layer. A higher stability of Bi in Bi_2O_2 layer than that in the perovskite-like layer is also implied, which is supported by the report of N. Sugita *et al.*[16] about the formation of Bi_2O_2 layer at lower temperature than that required for the perovskite-like layers containing TiO_6 octahedra.

From XRD, the degree of *c*-orientation, as shown in Fig. 3, increase continuously with increase of Bi excess from 0-4 mol.% and reaches a saturation in 4-7 mol.% but the film at 7 mol.% is not phase-pure BIT. Shulman *et al.*[17] suggests that the *c*-orientation development of BIT is very much related to the Bi_2O_2 layer.[18] From Fig. 4 and 5, the rigid layer mode (which we discussed previously and is largely due to the vibration of the Bi in Bi_2O_2 layer) remains unchanged at 59.22 cm^{-1} for 0-6 mol.% Bi excess; implying the Bi_2O_2 layer is intact in all Bi-deficient and Bi-saturated films having different preferred film orientations. Thus the preferred *c*-orientation, from our previous discussion of rigid layer modes, is not mainly due to the existence of Bi_2O_2 layer. The following discussion of the higher frequency modes may give us a clue to what mostly causes the preferred *c*-orientation in the BIT films.

Raman modes from ~ 121.6 cm^{-1} to the one just above 600 cm^{-1} are all corresponding to the vibrations inside the perovskite-like slab. The downshift of the mode at 121.6 cm^{-1} together with the shoulder peaks with the increase of Bi excess can be attributed to the increase of the distortion of TiO_6 octahedron.[19] A mode at ~ 200 cm^{-1} is due to the x(y) axis vibration of oxygen atoms, which is broadened with a decrease in intensity with increase of Ca content doped into the A site in the perovskite layer of $(Sr_{1-x}Ca_x)Bi_2Ta_2O_9$.[20] In our case, this mode at 194 cm^{-1} is highly overdamped with increase of Bi excess, implying that the excess Bi is incorporated into the A-site of the perovskite slab.

Considering the internal modes of TiO_6 octahedron, according to Ref. [13], the internal modes at ~ 229 cm^{-1} and ~ 267.8 cm^{-1} are for internal angle bending vibration; the one at ~ 335.2 cm^{-1} is for a combination of stretching and bending. The modes at ~ 267.8 cm^{-1} and 335.2 cm^{-1} are two characteristic peaks of BIT which persist into 2 mol.% Bi excess film. However, the mode at ~ 229.3 cm^{-1} emerging in 2 mol.% Bi excess film is unobserved in 0 mol.% Bi excess spectrum, and this is due to the aforementioned small grain size induced symmetry breaking effect. The mode at ~ 535.5 cm^{-1} with its shoulder at 2 mol.% Bi excess spectrum are also corresponding to the distortion of TiO_6 octahedron, and this mode becomes a single mode in 0 mol.% Bi excess indicating a symmetry breaking.

The internal mode at 616.14 cm^{-1}, which originates from the stretching of the octahedral O—Ti—O—Ti—O—Ti—O chain between two Bi_2O_2 layers, is well defined and remains unchanged in 2-6 mol.% Bi excess spectra but it becomes a broad band in 0 mol.% Bi excess spectrum. By taking average of the 'plateau' of this "overdamped" mode, its Raman shift is estimated to be 616.14cm^{-1}. Thus, this mode remains unchanged in all films, implying that the B site (Ti site) in the perovskite-like slab does not participate in the excess Bi incorporation process.

SUMMARY

From SEM images, the grain size and densification of BIT films increases with excess Bi content. This indicates that the grain growth of the film is related to the excess Bi content and liquid phase sintering occurs at 6 mol.% excess Bi BIT films. From the XRD data, the *c*-orientation of the BIT films increases from 0-4 mol.% and is well developed when Bi excess reaches between 4 and 7 mol.%. From the analysis of the Raman spectra, the Bi_2O_2 layer remains unchanged in the Bi-deficient (~ 0-2 mol.% Bi excess) as well as stoichiometric (~ 4-6 mol.% Bi excess) films; implying that during film growth, Bi first goes to the Bi_2O_2 layer in which the Bi site has higher stability than that in the perovskite-like slab. Raman modes indicate that with increasing Bi content, the excess Bi is incorporated into the A site in the perovskite-like slab, enhancing the distortion of TiO_6 octahedra, which may cause the increase of *c*-orientation of the films. To sum up, this study implies that a decrease of the distortion of TiO_6 octahedra by doping with a smaller ion like La is possible and is important for improving the polarization in the *c*-direction of BIT films. Also our study implies a higher stability of Bi in Bi_2O_2 layer than that of the Bi in perovskite-like slab, which explains why dopants (like La and Ca) incorporate into the A site in perovskite-like slab in BIT and SBT.

ACKNOWLEDGEMENT

We gratefully acknowledge Prof. Punit Boolchand (Dept. ECECS of UC) for his kind helps and generous sharing of Raman instruments. This work was partly supported by a NSF Grant (NSF-ECS 9612122).

REFERENCES

[1]C.F. Pulvari and A.S. de la Paz, "Phenomenological theory of polarization reversal in ferroelectric $Bi_4Ti_3O_{12}$ single crystals," *J. Appl. Phys.*, **37** [4] 1754-1763 (1966)

[2]S.E. Cummins and L.E. Cross, "Electrical and optical properties of ferroelectric $Bi_4Ti_3O_{12}$ single crystals," *J. Appl. Phys.*, **39** [5] 2268-2274 (1968)

[3]T. Kijima, "Relationship between Bi/Ti composition ratio and O_2 concentration for orientation control of MOCVD-grown $Bi_4Ti_3O_{12}$ thin films," *Electronics and Communications in Japan (Part 2)*, **84** [10] 49-58 (2001)

[4]J.T. Dawley, R. Radspinner, B.J.J. Zelinski and D.R. Uhlmann, "Sol-gel derived bismuth titanate thin films with c-axis orientation," *J. Sol-gel Sci. Technol.*, **20** [1] 85-93 (2001)

[5]R.C. Buchanan, R. Palan, A. Ghaffari, K. Tran and J.E. Sundeen, "Orientation effects on polarization and pyroelectric properties of ferroelectric thin films on Si," *J. Eur. Cer. Soc.*, **21** 1577-1580 (2001)

[6]P.R. Graves, G. Hua, S. Myhra and J.G. Thompson, "The Raman modes of the Aurivillius phases: temperature and polarization dependence," *J. Solid State Chem.*, **114** [1] 112-122 (1995)

[7]H. Idink, V. Srikanth, W.B. White and E.C. Subbarao, "Raman study of low temperature phase transitions in bismuth titanate, $Bi_4Ti_3O_{12}$," *J. Appl. Phys.*, **76** [3] 1819-1823 (1994)

[8]G.A. Kourouklis, A. Jayaraman and L.G. Van Uitert, "Pressure dependence of the Raman-active modes in $Bi_4Ti_3O_{12}$," *Materials Letters*, **5** [3] 116-119 (1987)

[9]C.Y. Yau, R. Palan, T. Khang and R.C. Buchanan, (unpublished)

[10]J. Kim, S.H. Kim, J.-P. Kim, Y.-H. Hwang and M.S. Jang, "A study of grain size dependent ferroelectric properties of annealed amorphous $Bi_4Ti_3O_{12}$," *J. Korean Phys. Soc.*, **35** [5] S1465-S1468 (1999)

[11]S. Kojima and S. Shimada, "Soft mode spectroscopy of bismuth titanate single crystals," *Physica B*, **219** & **220** 617-619 (1996)

[12]R. Zallen and M. Slade, "Rigid-layer modes in chalcogenide crystals," *Phys. Rev. B*, **9** [4] 1627-1637 (1974)

[13]S. Kojima, R. Imaizum, S. Hamazaki, and M. Takashige, "Raman scattering study of bismuth layer-structure ferroelectrics," *Jpn. J. Appl. Phys.*, **33** Part 1 [9B] 5559-5564 (1994)

[14]R. Zallen, M.L. Slade and A.T. Ward, "Lattice vibrations and interlayer interactions in crystalline As_2S_3 and As_2Se_3," *Phys. Rev. B*, **3** [12] 4257-4273 (1971)

[15]C.L. Fu and A.J. Freeman, "Optic breathing mode, resonant charge fluctuations, and high-T_c superconductivity in the layered perovskites," *Phys. Rev. B*, **35** [16] 8861-8864 (1987)

[16]N. Sugita, E. Tokumitsu, M. Osada and M. Kakihana, "*In situ* Raman spectroscopy observation of crystallization process of sol-gel derived $Bi_{4-x}La_xTi_3O_{12}$ films," *Jpn. J. Appl. Phys.*, **42** Part 2 [8A] L944-L945 (2003)

[17]H.S. Shulman, D. Damjanovic and N. Setter, "Niobium doping and dielectric anomalies in bismuth titanate," *J. Am. Ceram. Soc.*, **83** [3] 528-532 (2000)

[18]H.-S. Gu, A.-X. Kuang, S.-M. Wang, D.-H. Bao, L.-S. Wang, J.-S. Liu and X.-J. Li, "Synthesis and ferroelectric properties of *c*-axis oriented $Bi_4Ti_3O_{12}$ thin films by sol-gel process on platinum coated silicon," *Appl. Phys. Lett.*, **68** [9] 1209-1210 (1996)

[19]J.F. Meng, P.S. Dobal, R.S. Katiyar and G.T. Zou, "Optical phonon modes and phase transition in the $Bi_4Ge_{3-x}Ti_xO_{12}$ ceramic system," *J. Raman Spectrosc.*, **29** [12] 1003-1008 (1998)

[20]R.R. Das, W. Perez, R.S. Katiyar and A.S. Bhalla, "Structural investigation of $(Sr_{1-x}Ca_x)Bi_2Ta_2O_9$ ceramics," *J. Raman Spectrosc.*, **33** [4] 219-222 (2002)

HIGH DIELECTRIC TUNABILITY FERROELECTRIC (Pb,Sr)TIO_3 THIN FILMS FOR ROOM TEMPERATURE TUNABLE MICROWAVE DEVICES

S. W. Liu, Y. Lin,[*] J. Weaver, W. Donner, X. Chen,[†] C. L. Chen,[‡] H. D. Lee, and W. K. Chu
The Texas Center for Superconductivity and Department of Physics
University of Houston
Houston, Texas 77204

J. C. Jiang, and E. I. Meletis
Department of Mechanical Engineering
Louisiana State University, Banta Rouge, LA

A. Bhalla
Materials Research Laboratory
Pennsylvania State University, University Park
Pennsylvania 16802

ABSTRACT

Ferroelectric (Pb,Sr)TiO_3 thin films were epitaxially grown on (001) $LaAlO_3$ by using pulsed laser deposition. Microstructural characterizations with x-ray diffraction and transmission electron microscopy indicate that the as-grown films have excellent single crystalline quality and a $(001)_{PSTO}//(001)_{LAO}$ and $[100]_{PSTO}//[100]_{LAO}$ interface relationship. Dielectric property measurements reveal that the as-films have a very high dielectric constant value of 3100 and a very large dielectric tunability of 48 % at 40V/cm at room temperature. These excellent results suggest that the highly epitaxial ferroelectric (Pb,Sr)TiO_3 thin films can be developed for room temperature tunable microwave elements in wireless communication applications

INTRODUCTION

Ferroelectric materials with room temperature high dielectric constant, low dielectric loss, and large electrical tunability have extensively been studied in the past several years because of their potential in high frequency tunable microwave devices. Many research efforts have been focused on the ferroelectric (Ba,Sr)TiO_3 (BSTO) systems.[1] Remarkable progress has been achieved in highly epitaxial growth of ferroelectric BSTO thin films with high dielectric constant, low dielectric loss, and large dielectric tunability.[2,3,4,5] Various room temperature tunable microwave devices, such as microwave phase shifters, have been developed from the highly epitaxial BSTO thin films. However, the relatively high dielectric insertion loss, multiphase interaction, and high frequency soften mode have prevented the practical applications of ferroelectric thin films in high frequency tunable wireless communication in spite of many efforts in the past few years to improve the dielectric properties.[1] New materials are necessary for enhancing the dielectric properties and for convalescing the device applications. Recent research[6,7] revealed that ferroelectric (Pb,Sr)TiO_3 (PSTO) has extremely high dielectric tunability of 70% under 20 kV/cm at 10 kHz and room temperature with very low dielectric loss value of 0.001. These excellent dielectric properties suggest that the PSTO could have super-

[*] With Los Alamos National Laboratory, Los Alamos, NM 87545
[†] On leaving for the Texas A & M University, Department of Physics, TX
[‡] Correspondence should be addressed at clchen@uh.edu

passing properties to ferroelectric BSTO for developing high frequency room temperature microwave elements. Also, unlike the ferroelectric BSTO which possesses three phase transitions, PSTO only has one phase transition from cubic to tetragonal. This can avoid the microphase or precipitation formation in the films that alter the dielectric properties of the PSTO films.

On the other hand, the physical properties of PSTO are highly dependent upon the composition ratio between the Pb and Sr in the compound. Earlier studies indicated that the dielectric constant can reach to 15000 and the Curie temperature varies from 5 K to 770 K.[8] The dielectric constant and Curie temperature for a compound with ratio of 35:65 (Pb:Sr) is near 10000 and near 300 K, respectively.[8] Also, the unit cell volume remains unchanged for Pb content less than 50 %, which offers a great opportunity to improve the performance of high frequency tunable microwave elements.[8,9] Furthermore, it should be noted that decreasing the giant splitting of longitudinal and transverse optical phonons in ferroelectric PSTO suggests that this compound might prevail over the soft mode in the BSTO system. Theoretical studies reveal that the chemical bonding of Pb-O in PSTO is covalent rather than ionic as for the BSTO system.[10,11,12] This can be further manifested from the fact that the Pb-O interaction is more hybridized than the Ba-O interaction resulting in decreasing longitudinal-transverse splitting phenomena in the PSTO system rather than increasing splitting observed in BSTO.

EPITAXIAL GROWTH AND STRUCTURAL CHARACTERIZATION OF PSTO FILM

Synthesis of PSTO thin films by sol gel synthesis[13] and rf-sputtering technique[14] has recently been attempted to explore the advantages of this material. The results indicate that the PSTO thin films have very large dielectric tunability but with much higher applied electric field and larger dielectric loss value than its bulk material, probably due to its nonstoichiometry and growth defects. To fully understand the physical properties of PSTO thin films and to explore its advantage to the high frequency tunable microwave element applications, we have synthesized and investigated the PSTO thin films on (001) $LaAlO_3$ (LAO) by pulsed laser deposition. We were able to achieve excellent epitaxial quality and dielectric tunability with very low dielectric loss. The present results suggest that the highly epitaxial PSTO thin films can be developed for tunable microwave elements.

A LAO substrate was selected to epitaxially grow the PSTO thin films. The lattice misfit can be estimated to be about 2.5 % (a_{LAO} = 0.382 nm and a_{PSTO} = 0.392 nm). A target with the Pb/Sr ratio of 35:65 plus 20 at. % excess Pb was adopted for the film deposition to avoid the Pb deficiency because Pb is easily lost during deposition. Pulsed laser ablation was employed for the deposition of PSTO films. The irradiation energy density is 2.0J/cm^2 per pulsed in 4 Hz and the deposition rate was found to be near 7.0 nm/min. Microstructure characterizations by x-ray diffraction (XRD) and transmission electron microscopy (TEM) were performed to investigate the crystallinity and epitaxial quality. Dielectric property measurements were carried out with traditional interdigital technique.

Figure 1 is a θ-2θ scan along the surface normal from an x-ray diffractometer using Cu-radiation. The figure shows that only PSTO (001)-type reflections together with the corresponding reflections from the LAO substrate appeared in the pattern, indicating that the as-grown film is c-axis aligned. The high-resolution measurements ($\Delta\theta$=0.006°) of rocking scans in the (002) reflection show the tilt distribution of the film crystallites is extraordinarily narrow (only 0.054° of the full width half maximum), whereas the distribution of the crystallites' rotation has a width of 0.43°. From these observations and the widths of the radial scans, it is evident that the film consists of well aligned, 200 nm long columnar grains with an average diameter of 20 nm. These evidences indicate that PSTO films exhibit excellent epitaxy with very good single crystal quality. The film exhibits a slight tetragonal distortion with a c-axis length of 0.3914(1) nm and an a-axis of 0.3918(2) nm, respectively. There is a slight difference from the lattice parameters obtained from its PSTO polycrystalline powder which is cubic at room temperature as reported in the literature[9]. The distorted epitaxial film exhibits tetragonal symmetry probably resulting from thermal residual stress at the interface and the excellent epitaxial nature of the single crystalline PSTO film.

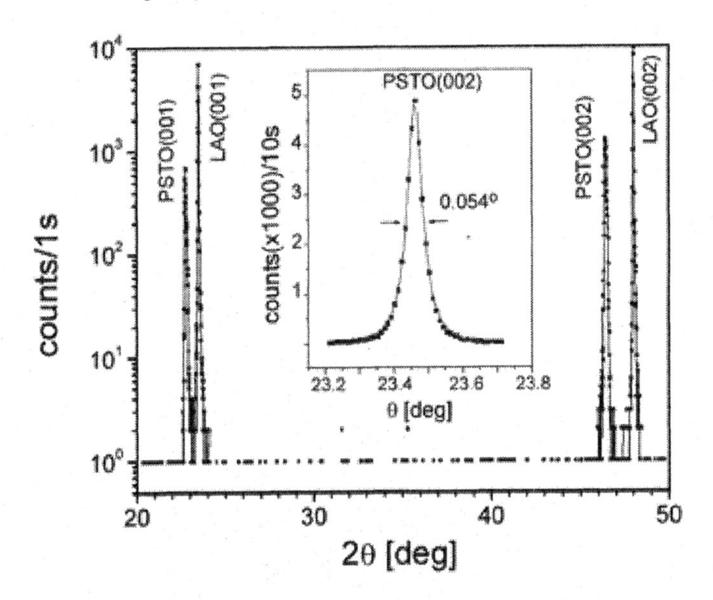

Figure 1, θ~2θ scan of x-ray diffraction showing that the as-grown PSTO films are c-axis oriented. Only (00l) reflections appeared in the diffraction pattern. The inset is a rocking curve measurement from the (002) PSTO reflection

To understand the epitaxial quality and interface structures, cross-sectional TEM studies have been conducted on a PSTO film deposited on (001) LAO using the same processing conditions as previously described. Fig. 2(a) is a low-magnification, bright field TEM image showing that the as-grown PSTO film has a sharp interface and good epitaxial behavior. The antidomain boundary formed directly at the interface between the PSTO film and the LAO substrates can be clearly seen in the image. This phenomenon is somewhat similar with the previous observations of BSTO films on (001) MgO[15,16] and (001) $LaAlO_3$.[17] Figs. 2(b) – (d) show the selected area electron diffraction (SAED) patterns taken along a [110] direction of the LAO substrate (b), the [110] direction of the PSTO film (c), and the interface covering both the substrate and the film (d), respectively. The film has excellent crystallinity that is evident from the sharp diffraction spots, as seen in Fig. 2 (c). Neither precipitate nor segregation was found in the films. The diffraction pattern at the interface is a simple superposition of the PSTO film and the LAO substrate. Again, the sharp electron diffraction spots with no satellites or broadening at the interface indicate that the film has excellent single-crystallinity. The interface relationship has been determined to be <100>PSTO // <100>LAO and (001)PSTO // (001)LAO, which agrees with the XRD studies. From the highly ordered electron diffraction spots, the lattice misfit can be estimated to be 4.0%, which is much larger than the value of 2.5% estimated from XRD and that calculated based on the lattice parameters from both PSTO and LAO unit cells. This difference could be due to the fact that the strain in TEM sample is one dimensional because the TEM sample is very thin along the interface whereas it is two dimensional in both x-ray samples and theoretical calculation. This phenomenon has been observed earlier in other systems.[18] The strain is usually relaxed by forming periodical edge dislocations at the interface between the film and the substrate. The high-resolution, cross-sectional TEM studies have demonstrated this epitaxial behavior, as seen in Fig. 2(e). The film has good epitaxial quality with a sharp interface structure. Edge dislocations were found to be formed over the entire interface with an average separation of ~ 26 lattice planes along the PSTO [110] direction, which also gives a lattice misfit in the range of 4.0 %, in good agreement with results from electron diffraction. This confirms that the PSTO film is of excellent single crystallinity. Neither a precipitate nor any other phase was present in the film or at the interface.

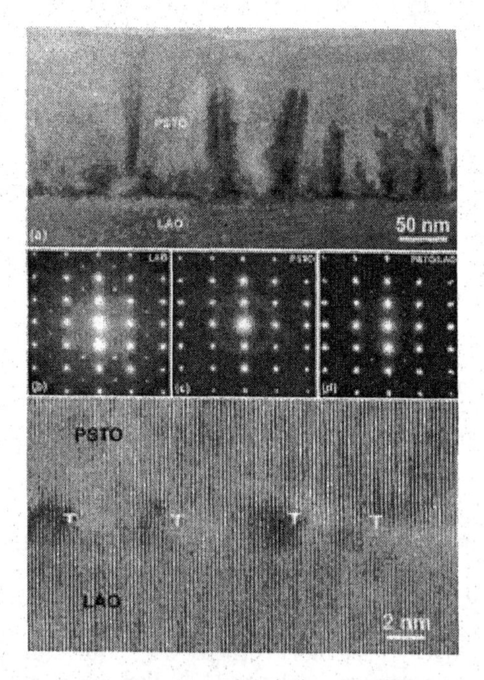

Figure 2, Cross-sectional TEM studies showing the epitaxial behavior of PSTO films on (001) LAO. (a) a low-magnification, bright field TEM image, (b) SAED from the LAO substrate, (c) SAED from the PSTO film, (d) SAED from the interface covering both PSTO film and LAO substrate, and (e) high resolution image showing the interface structure and epitaxial behavior, edge-dislocations can be seen at the interface

The crystallinity of the PSTO epitaxial films was investigated by using 1.8 MeV He ion RBS-channeling technique. Figure 3 shows the random and aligned spectra for a film deposited on (001) LAO substrate. The minimum yield taken behind the surface peaks is about 3.75%, indicating the excellent crystallinity of the PSTO films.

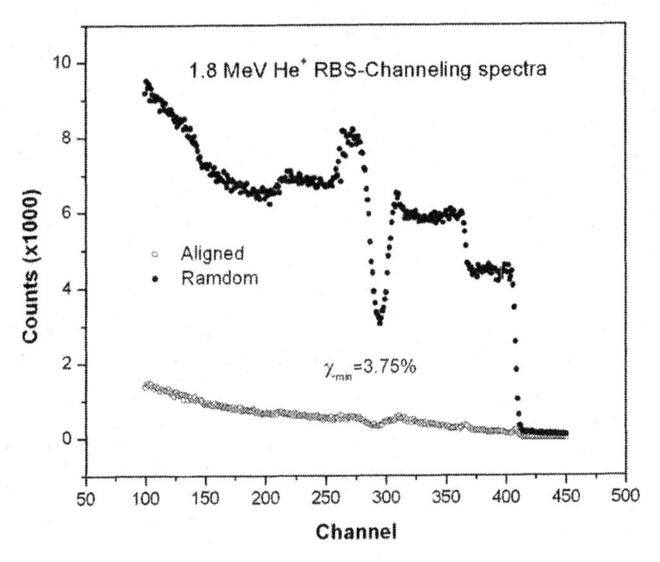

Figure 3, 1.8 MeV He ion channeling random and aligned spectra taken on an as-epitaxial PSTO film on (001)LAO substrate.

DIELECTRIC PROPERTIES

The dielectric property measurements performed by the traditional interdigital capacitor technique show that the room temperature dielectric constant and the dielectric loss tangent for a 200 nm thick film are ~ 3100 and 0.008, respectively, at 1 MHz (Figure 4). A large tunability of as much as 48% at 40 kV/cm (near saturated), or 34% at 20 kV/cm (unsaturated), has been achieved from the as-grown films. Thus, the dielectric properties of the highly epitaxial ferroelectric PSTO thin films are much better than the films prepared by other techniques. Also, this dielectric tunability value is somewhat similar to BSTO thin films on LAO[2] suggesting that epitaxial quality and dielectric properties of PSTO thin films are good for developing high frequency tunable microwave elements operating at room temperature. The higher frequency measurements and the developments of tunable microwave devices will be carried out at the Naval Research Laboratory and the Airforce Research Center and will be reported at a later time.

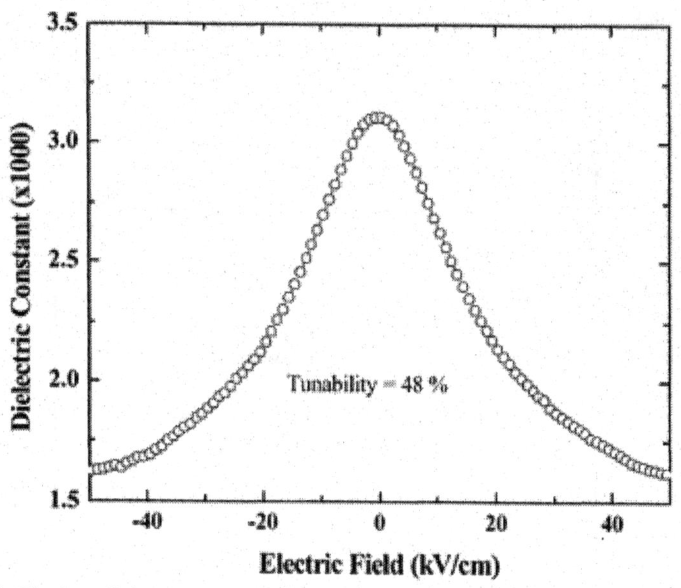

Figure 4, The interdigital dielectric property measurement showing the film have a very large dielectric tunability value of 48% at room temperature

SUMMARY

Dielectric PSTO thin films have been epitaxially grown on (001) LAO by pulsed laser deposition. Microstructure studies reveal that the as-grown films have excellent single crystallinity and epitaxial behavior. The interface relationship was determined to be [100]PSTO // [100]LAO and (001)PSTO // (001)LAO. Excellent room temperature dielectric properties are evident from an extremely high dielectric constant of 3100, a low loss tangent of 0.008, and a tunability of 48% at 40 kV/cm, suggesting that tunable microwave devices can be developed from the highly epitaxial PSTO thin films.

The work is partially supported by the State of Texas through the Texas Center for Superconductivity and Advanced Materials at the University of Houston, the National Science Foundation (NSF), and the Board of Regents of the State of Louisiana under the contract No. NSF/LEQSF (2001-04) RII-03.

REFERENCES

[1]For example, Materials Issues for Tunable RF and Microwave Devices III, edited by S. C. Tidrow, J. S. Horwitz, X. Xi, J. Levy, [Mater. Res. Soc. Symp. Proc.720 (2002)]; Epitaxial Oxide Thin Films III, edited by D. G. Schlom, C. B. Eom, M. E. Hawley, C. M. Foster, and J. S. Speck [Mater. Res. Soc. Symp. Proc. 474, (1997)]; Ferroelectric Thin Films II, edited by A. I. Kingon, E. R. Myers, and B. A. Tuttle [Mater. Res. Soc. Symp. Proc. 243, (1991)].

[2]C. L. Chen, H. H. Feng, Z. Zhang, A. Brazdeikis, Z. J. Huang, W. K. Chu, C. W. Chu, F. A. Miranda, F. W. Van Keuls, R. R. Romanofsky, and Y. Liou, Appl. Phys. Lett. 75, 412 (1999); i. b., 78 (2001) 652.
[3]K. R. Carroll, J. M. Pond, D. B. Chrisey, J. S. Horwitz, and R. E. Leuchtner, Appl. Phys. Lett. 62, 1845 (1993); i.b, 63, 1292 (1993).
[4]Q. X. Jia, J. R. Groves, P. Arendt, Y. Fan, A. T. Findikoglu, S. R. Foltyn, H. Jiang, and F. A. Miranda, Appl. Phys. Lett. 74,1564 (1999).
[5]Z.-G. Ban et al., J. Appl. Phys. 93, 504 (2003)
[6]Somiya Y, Bhalla AS, Cross LE, INTERNAT. J. INORG. MAT., 3, 709 (2001).
[7]Xing XR, Chen J, Deng JX, Liu GR, J. ALLOYS AND COMP., 360, 286 (2003).
[8]K. -H. Hellwege and A. M. Hellwege, "Ferroelectrics and Related Substances," subvolume a: Oxides, in Landolt-Bornstern Numerical Data and Functional Relationships in Science and Technology, edited by K. -H. Hellwege (Springer, Berlin, Heidelberg, and New York, 1981), Vol. 16.
[9]Kuo SY, Li CT, Hsieh WF, Appl. Phys. Lett., 81, 3019 (2002).
[10]R. E. Cohen, Nature (London), 358, 136 (1992).
[11]W. Zhong, R. D. King-Smith, and D. Vanderdilt, Phys. Rev. Lett., 72, 3618 (1994).
[12]Y. Kuroiwa, S. Aoyagi, A. Sawada, J. Harada, E. Nishibori, M. Takata, and M. Sakata, Phys. Rev. Lett., 87, 217601 (2001).
[13]Jain M, Majumder SB, Guo R, R, Bhalla AS, Katiyar RS, MAT. LETT., 56, 692-697 (2002).
[14]Karaki T, Du J, Fujii T, Adachi M, JAP. J. Appl. Phys. (PART 1) 41, 6761(2002).
[15]C. L. Chen, J. Shen, S. Y. Chen, G. P. Luo, C. W. Chu, F. A. Miranda, F. W. Van Keuls, J. C. Jiang, E.I. Meletis, and H. Chang, Appl. Phys. Lett., 78, 652 (2001).
[16]J.C. Jiang, Y. Lin, C.L. Chen, C.W. Chu and E.I. Meletis, J. Appl. Phys., 91 3188 (2002).
[17]H-J. Gao, C. L. Chen, B. Rafferty, S.J. Pennycook, G.P. Luo, and C.W. Chu, Appl. Phys. Lett., 75, 2542 (1999).
[18]X.Q. Pan, J.C. Jiang, W. Tian, Q. Gan, R.A. Rao and C.B. Eom, J. Appl. Phys., 86 (1999) 4188.

FABRICATION OF SELF-ASSEMBLED NANOSTRUCTURES OF MICROWAVE DIELECTRICS

S. Bhattacharyya, M. Jain, N. Karan and Ram S. Katiyar*
Department of Physics, University of Puerto Rico, San Juan, PR 00931,USA

ABSTRACT

Nanostructures of microwave dielectric (Ba,Sr)TiO_3 were prepared by a chemical solution decomposition route and a template assisted sol-gel route respectively. The two dimensional nanostructures exhibited a uniform and nearly monodispersed size, which was a strong function of the substrate. There seemed to be a mixture of particles with different orientations that was present in the thicker films also. The three-dimensional self-assembly showed interesting structures with promising optical characteristics.

INTRODUCTION

The quest for understanding nano particle processing methods is a natural consequence of the extensive activities carried out in the last decade that resulted in decreasing the size of electronic devices[1,2]. At present there are many ongoing research activities to explore the most rational and reproducible means of fabricating nano-patterns in a controlled manner[3,4,5]. Ferroelectrics are known to be a special class of functional materials that are promising for applications such as in high-density memory devices, high capacity antenna arrays, miniaturized sensors and transducers[6,7,8]. It is therefore highly relevant to the scientific community to fabricate and study ferroelectric materials at the nano-scale level. Previously, there have been some successful attempts to fabricate isolated ferroelectric PZT nanostructures self assembled at high temperature[9,10,11]. However, the positional tailoring of nano-dots remained to be a challenging task in the self-assembly process. It was also to be seen whether a high temperature refractory material would also undergo such micro-structural evolution.

In this paper, we are presenting the results of our attempts to fabricate nanostructures of microwave active materials such as BST using "self-assembled" "bottom up" approach. Although getting a regular arrangement of these particles on a solid surface is still in its infancy, it is possible to get isolated nano dots with reasonably uniform sizes located at certain specific places. The role of the substrate has been explored with a future target of inducing a spontaneous ordering of nano dots on a given surface. On the other hand, three-dimensional self-assembly of a template layer was explored to make periodic ferroelectric nanostructures which have been characterized using of Raman spectroscopy and optical scattering measurements.

EXPERIMENTAL

Ultra-thin films of (Ba,Sr)TiO_3 with several Ba/Sr ratios were deposited using a metal-organic decomposition route. Stoichiometric quantities of barium acetate, strontium acetate and titanium iso-propoxide were added in acetic acid in order to make the solution for BST films. The sol was diluted to 0.33 M/lit by acetic acid, which was considered to be the initial dilution. The sol was further diluted to several concentrations up to 1:20 in steps. The reason for choosing this dilution was, the parent sol gave a thickness of 50 nm after the first coating. According to some

* Corresponding author. Email: rkatiyar@speclab.cnnet.clu.edu

reports, the initial film should be less than 40 nm thick for appropriate microstructural evolution[12]. Films were deposited on $SrTiO_3$, MgO, $LaAlO_3$, and Nb doped $SrTiO_3$ substrates. All substrates were [100] cut, and polished to a roughness less than 1 nm (purchased from MTI crystals, CO). After deposition, the films were pyrolized at 400 °C to remove the organic part. The pyrolized films were then heated at 1100 °C for one hour. Films were morphologically characterized by an atomic force microscope (Digital Instruments, Nanoscope IIIE).

For the 3D self-assembled crystal preparation, a 10% water dispersion of polystyrene spheres (446 nm, 10 wt% in water, from Aldrich) was used. We have diluted it further to 2.5 wt percent by mixing with water, and added a surfactant (TRITON X-1000, from Alfa-Aesar) with the diluting water at a weight ratio of 1:500. A few drops of the final dispersion was dropped on a quartz substrate and allowed to evaporate in vacuum at 60 °C. After the template was dried, one drop of BST solution was dropped on it and was spun at a low speed (~1000 rpm). The drop of BST sol was then kept in a high humidity environment for two days, that enabled the sol to infiltrate slowly inside the voids of the template and get hydrolyzed. Once the sol was completely hydrolyzed, the template was slowly heated to a temperature of 600 °C and was kept for one hour.

RESULTS AND DISCUSSION

It was reported in the literature that the perovskite oxides would have a natural tendency to follow the orientation of the substrate if the substrate is also another perovskite material[13]. We have tested this on thicker films of BST grown on $SrTiO_3$ that is shown in the Figure 1. The following film was approximately 400 nm thick and was annealed at the same temperature (1100 °C/60 min.) at which the subsequent thinner films were grown. It was quite evident from the figure that the film was highly oriented on the lattice matched $SrTiO_3$ substrate. However, there were some other weak peaks observed in the sample, which were almost negligible in comparison to the oriented part of the film. These small peaks have been detected in the logarithmic scale of intensity as shown in the inset of figure 1.

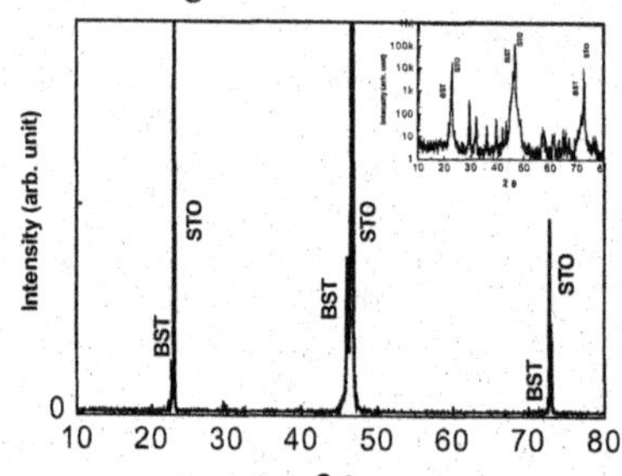

Figure 1. XRD of highly <001>oriented BST film on <001> oriented $SrTiO_3$ substrate (the inset shows the intensity in the log plot)

We expected that the orientation relationship between the substrate and the film would be maintained in case of the thinner films as well. Due to a reduction of the total X-ray irradiated volume in the ultra-thin samples, it was however, not possible to get any direct XRD signals from them. Therefore, We had to extrapolate the results of the thicker films to the case of thinner films. It is established that the lattice mismatch leads to an increase in the elastic energy of the over-layer when it is forced to elongate or compress due to the substrate clamping. The initial few layers during the growth process will strain by themselves to fit with the substrate lattice. However, if the film is thick enough, then straining the entire film will cost much more energy than the film relaxing the strain by interrupting the periodicity at some places. This can happen by the creation of grain boundaries, i.e., by forming a polycrystalline sample. We therefore believed that, if the films in our studies were made thinner, it would have an even better orientational relationship with the substrate.

Figure 2 shows the effect of annealing the film with an initial sol dilution of 1:2 (initial thickness ~ 25 nm) with respect to the original film. The notation 1:2 stands for the

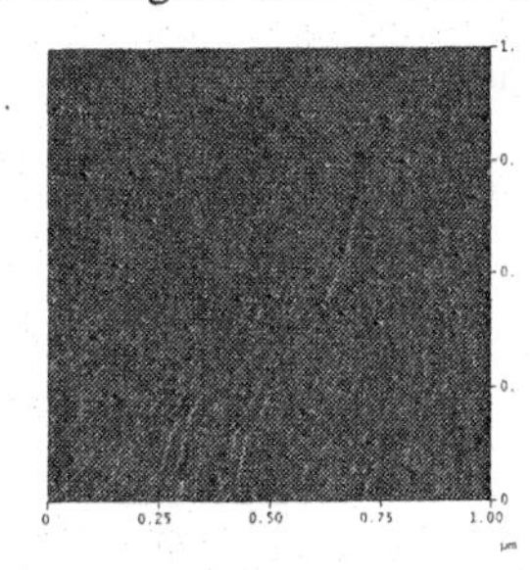

Fig. 2: the annealed continuous film made with the 1:2 diluted solution

volume ratio of the parent sol to acetic acid, which was used as the solvent. It was seen to have undergone a significant surface reconstruction during annealing. The surface reconstruction was evident form the wavy patterns that formed after the heat treatment. The film remained continuous all over the substrate.

Figure 3. The 1:4 diluted sol, single coat/ after annealing. A small region has been zoomed in to show the small particles

Figure 3 shows the effect of diluting the sol to a factor of 4 (1:4). It was seen that the annealed film had developed some well-defined cracks with their sidewalls aligned approximately by 45 degrees with the substrate edges, and the sidewalls of these pit holes intersecting at 90 degrees. There were no site selection on the location of these pit holes, but they were all parallel to each other, with a distribution in their sizes. The regions where the pit holes had developed were all topographically alike, and this led to the conclusion that the entire film was a single crystal, save from some small particles which were spread all through out the film but did not look like an integral part of the rest of the (single crystalline) film. The diameter of these particles was of the order of 25-30 nm, and there were no external features in their topography. They were mechanically very stable, since they remained at their positions even after repeated AFM scans, leading to a belief that they are the parts of the film. Considering their sizes in comparison with the single crystalline region of the film, we have come to a conclusion that these are the polycrystalline parts of the film, i.e., the off oriented grains which were present even in a thick film as evident from figure 1. More systematic analysis with the help of TEM and XPS measurements are underway and high resolution XRD measurements are in our plan.

This trend continued to the higher dilution levels. It was seen that the pit holes grew laterally as the initial thickness of the film was monotonically decreased and they finally joined together completely isolating the solid regions of the film from each other. In other word, the film was completely split into islands. The size of the islands was also reduced when the thickness was decreased further. This trend has been indicated in the figure 4. The size of the islands also shrank upon further dilution of the initial sol to 1:20, and the final size was about 90-100 nm, as shown in figure 5. It was seen that the substrate exhibited a major surface reconstruction during annealing, and the formation of

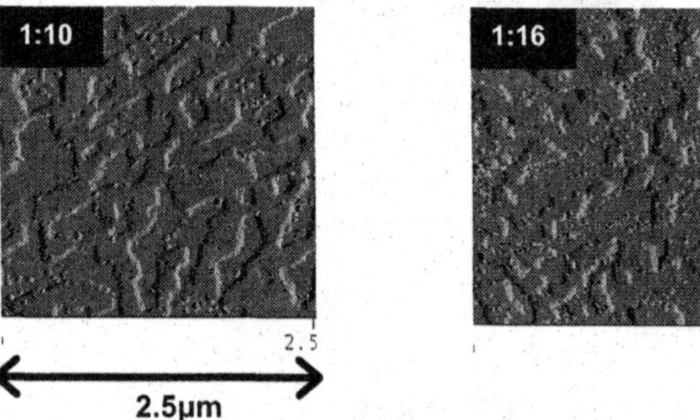

Figure 4. The effect of further dilution (the dilution levels are indicated in the figure).

the nano islands were also guided by the microscopic defects. The small particles were situated on the step edges formed on the substrate. We believed that this was due to the formation of low

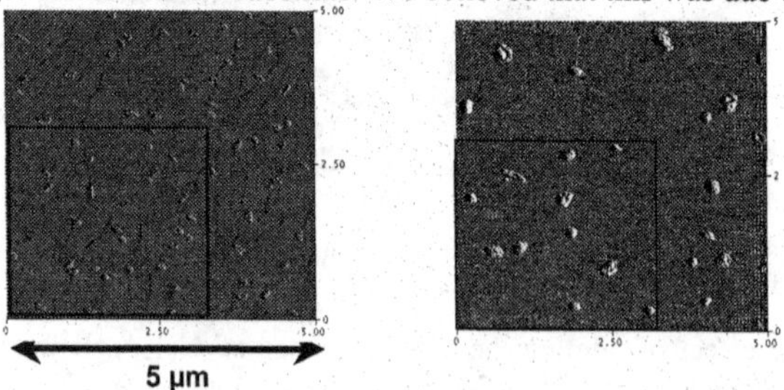

Figure 5. (a) Film that has been prepared at a spin speed of 5000 rpm (b) Film that has been prepared at a speed of 6000 rpm.

energy sites at these locations. On the other hand, we have performed the same experiment with a substrate having a few polishing defects prior to the deposition, and found that these macroscopic defects had no role to play in the registration of the nano-dots to the substrate. Figure 5 also compares the effect of the decrease of the initial thickness beyond this limit. If the initial thickness was reduced further (than what was obtained from the 1:20 diluted sol), then the size of the particles did not change appreciably, instead, the material conservation was maintained by the reduction of the areal density of the particles. Therefore, it seems evident from our results that, the lowest size of the dots have a thermodynamic limit, which was probably governed by the film-substrate interface characteristics.

The fact that the density of nano-islands could be increased (or decreased) without substantially changing the particles size actually enhanced the possibility of stacking many proximity-free nanodots in a smaller region, which is necessary for future generation high density memory devices. The abovementioned observations were already established in lead based thin films having perovskite structure. It was pointed out by F. Lange *et. al.*[12], that, an ultra-thin continuous perovskite crystalline film would be unstable compared to a discontinuous film oriented along the same direction. The thinner the film would become, the more the instability would be which finally could lead to a complete nanostructuring. It was quite clear that, without an external force, no material would be energetically stable to remain at an extremely flattened state as a thin film. In the real case, the adhesion with the substrate serves as the source of additional energy that can force a thin film to remain in a highly deformed state. The thinner a film would become, the more will be the surface to volume ratio of the system, thus making it energetically unstable due to the increase in surface energy. However, when the wetting is good, the film would not gain much in energy by splitting at random places. In such cases, the breaking would be initiated only along some very low energy cleavage planes, which were the (111) plane in the perovskite lattice. In one word, on a lattice-matched substrate, a perovskite film would only break along the (111) plane rather than any other planes. This also explained the creation of pit holes in the initially continuous film when its entire exposed surface was along the (001) plane. When a tapered pit hole formed, it reduced some area from the exposed (001) surface, and created some new surface, which was along the (111) planes.

This theory explained the structural evolution in the single crystalline film, but did not hint on the final shape and microstructure of an ultra thin film that was in a polycrystalline state. For lattice-mismatched substrates, it is very likely that the film would grow in a polycrystalline structure. It was previously predicted that the polycrystalline films also developed micro-structural instability when heated to high temperature, and this was achieved by grain boundary grooving phenomenon[14]. In such a system, grain boundaries present in the polycrystalline samples would start deepening, exposing the bare substrate.

We have repeated the same experiment with different substrates such as MgO, and $LaAlO_3$. MgO would introduce an elongative strain due to a larger lattice parameter (4.20 Å) than BST, whereas $LaAlO_3$ would introduce a compressive strain due to smaller lattice parameter (3.82 Å). From normal experience one would expect that the film would easily buckle off when it is under a compressive strain but would sustain the same amount of elongative strain without changing the surface topography. The lattice mismatch would also be reflected in the interface energy between the two materials. The interface energy was expected to play a very crucial role, since, without the interface wetting, the film will try to acquire its minimum energy shape which would be a spherical solid droplet, or a well-defined single crystal when the surface energy is highly anisotropic. Therefore, it was logical to assume that LAO would lead to a better nano-structuring than a lattice matched substrate (like STO), and MgO also might still prefer to have a continuous coverage. However, when the films of the same thickness on two different substrates were annealed at the same temperature, we obtained exactly opposite results. Figure 6 highlights the effect of the substrate on micro-structural evolution. Contrary to the expected results, we

Figure 6. (a) Film grown with a 1:20 diluted sol on MgO substrate.
(b) Film grown with the same dilution on $LaAlO_3$ substrate. Scan size in both films are 5 μm.

have noticed that, the film, which was grown on MgO, formed larger globules (~250 nm) whereas the one grown on $LaAlO_3$ was still not well split at the same annealing temperature, which indicated that the film had a better wetting with $LaAlO_3$. This also indicated that the wetting was probably not determined by the lattice parameters alone, but also by the lattice type. $LaAlO_3$ had a perovskite lattice similar to the BST lattice itself, and MgO had a spinnel structure thus making it more unfavorable to stick to a perovskite lattice than a perovskite substrate can stick to a perovskite film. Since the adhesion on MgO was poor in films grown on MgO, it was the cohesive energy of the atoms within the film itself, which played the major role in the energetics, and therefore, the small droplets coalesced together to form a bigger globule. On the other hand, the film on LAO will try to stick to the substrate unless the thickness is made extremely small. The very fine undulations of the film on the LAO substrate indicated that, it would give rise to the formation of much finer particles than on MgO and STO.

Self-assembly in Three Dimensions: As the micro-structural instability proved to be a good way of two-dimensional nanostructuring, three-dimensional nanostructuring was done by the nano-templating method, which was already described in the experimental section. Here, as the template was soaked in the BST sol and was kept for some time, the voids of the template got

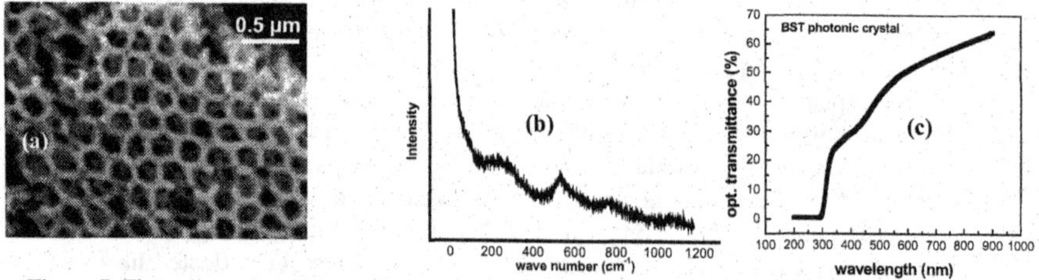

Figure 7. Three-dimensional mesoporous BST network prepared by self assembled templating. (a) SEM, (b) Raman spectra, (c) Optical transmission spectra

filled with the sol, which was hardened later, and these hardened regions remained stable when the basis template structure was removed by thermal etching. We have obtained the mesoporous BST with 250 nm pore size and a 30-40 nm thickness of the walls separating the pores. The three dimensional nature of the structure was evident from the figure 7. Since the major portion of the structure was hollow, the amount of material present was very smalml, and it was not possible to obtain any signals by X-ray diffraction. We have confirmed the phase formation through Raman spectroscopy that matched well with the bulk BST. We have also measured the optical transmittance property of this structure (which was expected to have photonic properties[15]), and found that the absorption showed a double knee. Since the structure was grown on a glass substrate, the second knee was believed to be due to the substrate whereas the first knee could be due to the Bragg reflecting sample with the artificial periodicity in the structure.

CONCLUSSIONS

Self-assembled structures were demonstrated in two and three dimensions. Perovskite BST was chosen as a model system and since, no properties, which were material specific, were used here, it expected that the phenomenon was general and could be applied to any material system, with a proper choice of the substrate. We were able to get the nano particles of sizes 90-100 nm, and still finer sizes appeared to be possible by choosing appropriate substrates. We were able to tailor the positions of these nano sots to a certain extent, through the substrate surface reconstruction. On the other hand, macroscopic defects were found to have negligible effects on the site selection of the nucleation of nano-dots. We have also demonstrated that below a certain size of the nano-dots, it was possible to change their areal density without affecting their size and shape any further. The lowest size achievable by this method had a limit determined by the substrate-film interface bonding. It was concluded that the micro-structural evolution in single crystalline films occurred by the formation of pit holes along the (111) planes, and in polycrystalline films on lattice mismatched substrates, the nanostructuring occurred through grain boundary grooving, and it was found that the cohesion among the nano particles made them coalesce to form larger particles. Three-dimensional nanostructuring was demonstrated by a templating method, and the interesting optical properties were indicated.

ACKNOWLEDGEMENTS

The authors would like to acknowledge the NSF-DMR (FY 2004) and NASA NCC-1034 for their financial support in this project.

REFERENCES

[1]"20 nm Resolution of Electron Lithography for the Nano-Device on Ultra-Thin SOI Film," Y.V. Nastaushev, T. Gavrilova, M. Kachanova, L. Nenasheva, V. Kolosanov, O.V. Naumova, V.P. Popov and A.L. Aseev, *Materials Science & Engineering C: Biomimetic Materials, Sensors & Systems,* **C19**, 189 (2002).

[2]"Ballistic & Tunneling GaAs Static Induction Transistors: Nano-Device for THz Electronics," *J. Nishizawa, P. Plotka and T. Kurabayashi, IEEE Transactions on Electron Devices,* **49**, 1102 (2002).

[3]"Bio-MEMS and Smart Nanostructures," V. Millar, C.I. Pakes, A. Cimmino, D. Brett, D.N. Jamieson, S.D. Prawer, C.J. Yang, B. Rout, R.P. McKinnon, A.S. Dzurak and R.G. Clark, *Proceedings of SPIE,* **4590,** 173 (2001).

[4]"The Role of Nucleation and Heteroepitaxial Process in Nanostructuring on Si," E.A. Guliants, C. Ji and W.A. Anderson, *Journal of Electronic Materials,* **31**, 466 (2002).

[5] "Well-ordered Arrays of Pyramid-Shaped Ferroelectric $BaTiO_3$ Nanostructures," Wenhui Ma, Catalin Harnagea, Dietrich Hesse, and Ulrich Gösele, *Applied Physics Letters,* **83(18),** 3770 (2003).

[6]"A Robust $8F^2$ Ferroelectric RAM Cell With Depletion Device (DeFeRAM)," G. Braun, H. Hoenigschmid, T. Schlager, and W. Weber, *Journal of Solid State. Circuits.* **35,** 691 (2000).

[7]"Embedded Self-Sensing Piezoelectric Active Sensors for On Line Structural Identfication," V. Giurgiutiu, and Z. An, *Journal of Vibration & Acoustics-Transactions of the ASME,* **124,** 116 (2002).

[8]"Sol-Gel Derived Grain Oirented BST Thin Films for Phase Shifter Applications," S.B. Majumder, M. Jain, A. Martinez, F.W. Van Keuls, F.A. Miranda, and R.S. Katiyar, *Journal of Applied Physics,* **90**, 896 (2001).

[9]"Ferroelectric Epitaxial Nanocrystals Obtained by a Self-Patterning Method," I. Szafraniak, C. Harnagea, R. Scholz, S. Bhattacharyya, D. Hesse, and M. Alexe, *Applied Physics Letters,* **83(11),** 2211 (2003).

[10]"The Contribution of Asymmetric Strain Fields in Epitaxial $Pb(Zr_{0.52},Ti_{0.48})O_3$ Nanoislands to Ferroelectric Size Effects," Ming-Wen Chu, Izabela Szafraniak, Roland Scholz, Dietrich Hesse, Marin Alexe, and Ulrich Gösele, *Materials Research Society Symposium Proceedings,* **784,** C1.3.1 (2004).

[11]"Piezoresponse Force Microscopy of Lead Titanate Nanograins Possibly Reaching The Limit of Ferroelectricity," A. Roelofs, T. Schneller, K. Szot, and R. Waser, *Applied Physics Letters,* **81**, 5231 (2002).

[12]"Microstructural instability in single-crystal thin films," F. A. Seifert, A. Vojta, J.S. Speck, and F.F. Lange, *Journal of Materials Research,* **11,** 1470 (1996).

[13]"Formation of Epitaxial $BaTiO_3/SrTiO_3$ Multilayers Grown on Nb-Doped $SrTiO_3$ (001) Substrates,"A. Visinoiu, R. Scholz, S. Chattopadhyay, M. Alexe, and D. Hesse, *Japanese Journal of Applied Physics,* **Part I, 41,** 6633 (2002).

[14]"Finite Element Analysis of the Splitting & Cylinderization Process of Damage Microcracks," Peizhen Huang, Zhonghua Li and Jun Sun, *Modelling & Simulations Materials Science and Engineering*, **9,** 193 (2001).

[15]"Ferroelectric Inverse Opals with Electrically Tunable Photonic Band Gap," Bo Li, Ji Zhou, Longtu Li, Xing Jun Wang, Xiao Han Liu, and Jian Zi, *Applied Physics Letters,* **83(23),** 4704 (2003).

EPITAXIAL PHASE SELECTION IN THE RARE EARTH MANGANITE SYSTEM

K. R. Balasubramanian, Arati A. Bagal, Oscar Castillo, Andrew J. Francis and Paul A. Salvador

Department of Materials Science and Engineering, Carnegie Mellon University
Pittsburgh, PA 15213

ABSTRACT

Manganites of the smaller rare-earths (*RE* = Ho, Er, Tm, Yb, Lu, Y and *Dy*), crystallizing in a hexagonal structure, belong to a class of materials known as the hexagonal ferroelectrics. The larger rare-earth manganites crystallize in an orthorhombic, perovskite-related form. We are interested in the epitaxial stabilization of the larger rare-earth manganites in the hexagonal structure to develop improved ferroelectrics. We will focus on the epitaxial stabilization of $DyMnO_3$, the crossover material, in both the hexagonal and orthorhombic forms. $DyMnO_3$ is easily stabilized in the orthorhombic structure, on $SrTiO_3$(111) from 700-900 °C. The phase pure hexagonal form grows on MgO(111) at 700 and 800°C, while at higher temperatures decomposition to the orthorhombic phase begins. The implications of this result are discussed with respect to the growth of phase-pure metastable hexagonal ferroelectrics, such as $GdMnO_3$ that is observed to grow as a multiphasic mixture on MgO(111) at all temperatures.

INTRODUCTION

The stable hexagonal $LuMnO_3$ structure is adopted by compounds with the formula $RE$$MnO_3$, where *RE* = Ho, Er, Tm, Yb, Lu and Y.[1] This structure is non-close-packed and layered, and the Mn^{3+} cations are 5-fold coordinated with the oxygen ions located at the corners of the trigonal bipyramids. The RE^{3+} ions are 7-coordinated with the oxygen (in edge-sharing, distorted, bicapped octahedra), and are located in between the trigonal bipyramids. $YMnO_3$ was one of the first compounds in this family to be discovered. Fujimura *et al*[2,3] have proposed the use of *h*-$YMnO_3$ thin films on Si for use in non-volatile RAM applications.

The stable perovskite $GdFeO_3$ structure is adopted by the $RE$$MnO_3$ compounds with *RE* = Tb, Gd, Eu, Sm, Nd, Pr, La.[4,5] Partial substitution of *RE* by Ca, Sr, Ba results in colossal magnetoresistance[6] near the spin-ordering temperatures of Mn ions; similar compositions are also used as electrode materials in SOFC applications.[7] The $GdFeO_3$ structure has been well described by Geller *et al.*[8] This perovskite phase is metastable for the smaller rare-earth manganites and has been obtained in the case of $YMnO_3$ by high pressure synthesis,[9] soft chemistry routes,[10] or epitaxial stabilization on perovskite substrates.[11] In the case of $DyMnO_3$, there has been some confusion in the literature on the stable modification. Some authors[10,12] claim that it is the perovskite, while others claim that it is the hexagonal form. Szabo[11] has shown that quenching from temperatures in excess of 1600 °C results in the hexagonal modification of $DyMnO_3$; otherwise the perovksite form results. Based on the observations of Szabo, that the hexagonal polymorph is the high temperature form of $DyMnO_3$, Bosak *et al*[12] concluded that $GdMnO_3$ has to be quenched from T > 2800 °C to obtain the hexagonal polymorph. They suggest epitaxial stabilization as a potential route to synthesizing the larger rare-earths in the hexagonal polymorphic structure.

We have been interested in the use of epitaxial stabilization to realize novel materials,[11,13-16] including the rare earth manganites. The confusion over the stable polymorph for $DyMnO_3$ stems, in part, from the fact that this compound exists near the crossover of stability for the two polymorphs. To address this issue we adapted experimental values to temperatures appropriate to film growth. Figure 1 depicts approximate free energy of formation of the compounds in the rare-earth manganite series, as a function of the rare-earth cation radius, at 800 °C. To create this plot, we used the values of the ionic radii for the rare-earth in 12-fold coordination, obtained from the data of Shannon.[17] The thermodynamic data was adapted from those given in Kamata *et al*,[18] which reported the experimentally determined free energies of formation of these compounds at 1200 °C. We used the data reported in Satoh *et al*,[19] which looked at the thermodynamic data as a function of temperature, to extrapolate the high temperature values to 800°C. For compounds that were not experimentally available, the free energy data was then linearly extrapolated with respect to the ionic radius to estimate the values for the metastable compounds. Based on these approximations, the stable form for $DyMnO_3$ is predicted to be the hexagonal form, although the difference in stability is small.

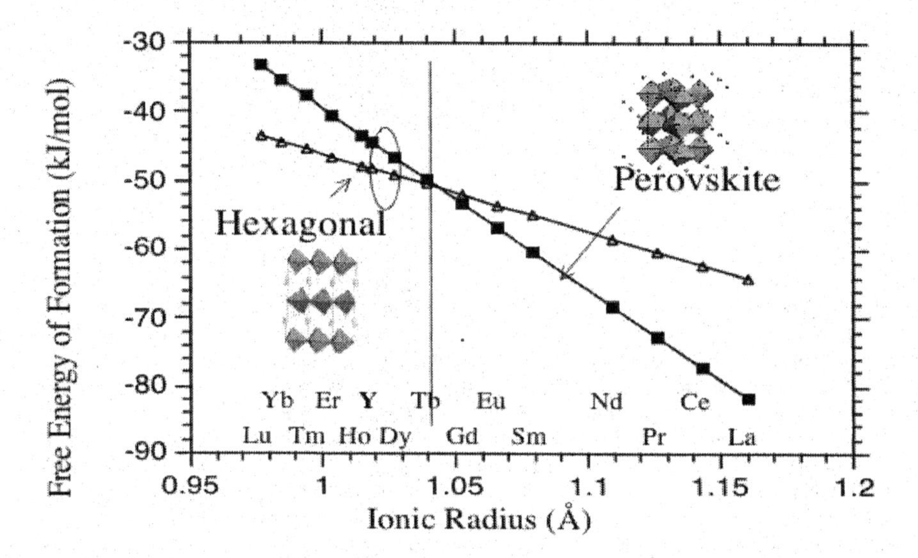

Figure 1. ΔG formation vs ionic radius of the *RE*MnO_3 compounds at 800 °C.

During thin film growth, several other energetic terms are important and can alter the phase stability of the growing film. The general expression for the change in free energy accompanying the formation of a film on a substrate will be given by,

$$\Delta G = V(\Delta G_v + w_s) + \Sigma A_i.\gamma_i. \qquad (1)$$

In this expression ΔG_v is the free energy change accompanying the formation of a phase and w_s is the strain energy term for the film arising from its interaction with the substrate; these terms scale with the volume. The γ_i terms are the interfacial energies accompanying the formation of the film, such as film-substrate or film-vapor interfaces; these terms scale with their surface area, A_i.

In this paper, we will be concerned with how growth on different substrate surfaces affects the polymorph stability of rare-earth manganese oxide phases near the crossover region seen in Fig. 1. Our aim is to better understand the experimentally observed influence of the different energetic terms, $\mathbf{\Delta G}$, $\mathbf{w}_s$, and γ_I, on phase selection.

EXPERIMENT

The oxide films were deposited on 0.5•m–thick substrates of commercially bought (Crystal GmbH, Berlin, Germany) MgO(111) and $SrTiO_3$(111) single crystals. The substrates were sectioned using a diamond wire saw into 5 mm x 5 mm pieces that were then ultrasonically cleaned using acetone and ethanol. Films of $DyMnO_3$ and $GdMnO_3$ were deposited on both these substrates via pulsed laser deposition, using a procedure described in detail elsewhere.[16] Briefly, a KrF excimer laser operated at a pulse rate of 3 Hz was focused to a laser energy density of • 2 J/cm^2 at the ceramic target surface. The substrates were positioned approximately 6 cm from the target and were heated to a temperature between 600 °C-900 °C during deposition, as prior research in the authors' laboratory had shown that crystalline $YMnO_3$ films were obtained in this temperature range. The dynamic chamber pressure was maintained at 5 mTorr O_2 during growth. Films were deposited for 1 hour to obtain a thickness of approximately 100 nm. The crystal structures of these films were determined by normal X-ray diffraction on a Philips X'Pert system system (Philips Analytical X-Ray B.V., The Netherlands). The epitaxial relationships of the thin films with the substrates were determined by phi scans using the same system.

RESULTS AND DISCUSSION

Films were deposited upon two surfaces that have a three-fold axis perpendicular to the surface plane, such that gross differences in interface symmetry could be ignored in this comparative investigation. On MgO(111), neither the hexagonal phase nor the perovskite phase have lattice-matched planes that would lead one to expect a coherent, low-energy interface to form. It is known that hexagonal $YMnO_3$ can be grown epitaxially upon MgO(111), forming a semi-coherent interface. Figure 2 depicts the X-ray diffraction scans of $DyMnO_3$ films deposited on MgO(111) at different temperatures. The peaks marked **m** originate from the substrate. At 600°C, no diffraction peaks are observed from $DyMnO_3$. Similar results have been found for other rare-earth manganese oxides; it is believed that the crystallization temperature is greater than 600°C for this family. At 700°C, weak peaks from the hexagonal phase (marked h-d(00*l*)) are observed. At 800°C, good quality hexagonal $DyMnO_3$ is observed; these patterns are reminiscent of epitaxial $YMnO_3$ on MgO(111).[3] Upon further increase in temperature, small peaks (marked *) from the orthorhombic perovskite phase are observed, implying that the hexagonal phase is competing with the orthorhombic phase at this temperature on MgO(111). $YMnO_3$, on the other hand, is stable on this surface even up until 900°C. Further investigations are required to understand better whether the decomposition is related to a bulk thermodynamic factor, a kinetic factor, or both.

The in-plane epitaxy of the film was determined by registering azimuthal ϕ-scans and by comparing the locations in ϕ-space of the $DyMnO_3${112} and MgO{200} reflections. The two phi scans, taken from the $DyMnO_3$ film deposited on a MgO(111) substrate at 800 °C, are overlayed in Figure 3. It should be emphasized that they are taken at different two-theta and psi (the sample tilt) angles (2θ, ψ = 22.87, 15.24, for $DyMnO_3${112} and 2θ, ψ = 42.98, 54.74, for

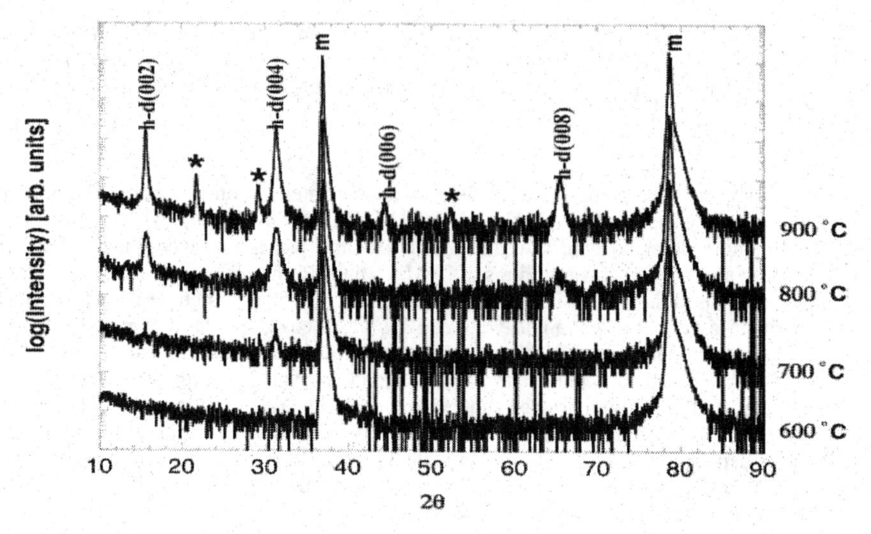

Figure 2. Normal θ–2θ scans of hexagonal $DyMnO_3$ films grown on MgO(111) substrates.

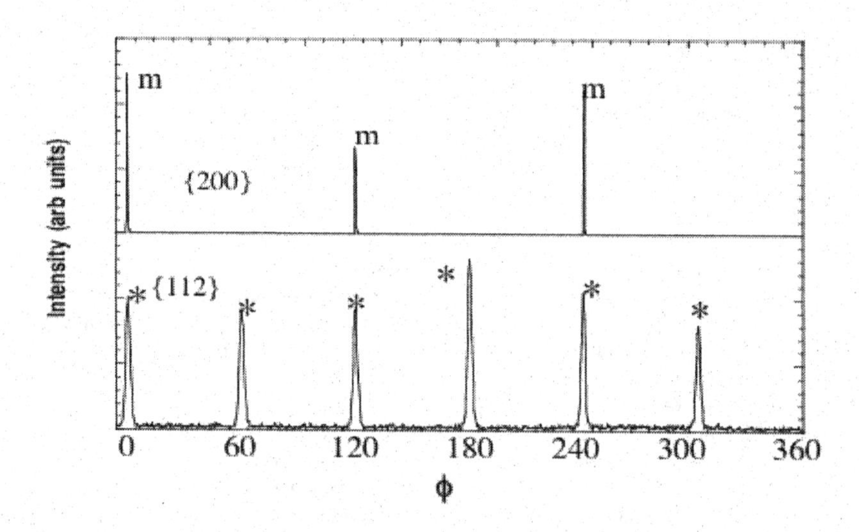

Figure 3. Phi scan of the $DyMnO_3$ film on MgO(111) substrate illustrating the epitaxial relationship between film and substrate.

MgO{200}). In the figure, the MgO peaks are labeled as m and the $DyMnO_3$ peaks with an asterisk. It is immediately obvious that observed six-fold intensity of the {112} $DyMnO_3$ suggests that there is a single hexagonal texture in-plane. The ϕ-scans indicate that the $DyMnO_3${112} peaks are exactly aligned with the MgO{200} peaks in ϕ-space. Taken together (Figs 2 and 3), these two diffraction scans give us an epitaxial relationship for $DyMnO_3$ that is $(001)_{DyMnO3}$••$(111)_{MgO}$ and $[1\bar{1}0]_{DyMnO3}$ •• $[1\bar{1}0]_{MgO}$. In other words, the in-plane arrangement is such that the densely-packed directions align themselves, which would result in a semi-coherent interface structure.

Although the above results do not answer the question of absolute polymorph stability (which is further discussed below), it is clear that thin film deposition allows for the facile synthesis of $DyMnO_3$ in the hexagonal structure. Normal powder synthesis techniques, such as that used to create the target, lead to orthorhombic $DyMnO_3$. The formation of the orthorhombic perovskite variation of $DyMnO_3$ as a thin film was achieved through epitaxial stabilization on a cubic perovskite substrate, $SrTiO_3$(111). Unlike the MgO(111) surface, the close-packed perovskite substrate surface is expected to form coherent (nearly lattice-matched) interfaces with the close-packed planes in the perovskite film. Figure 4 shows the XRD patterns of the $DyMnO_3$ films grown on $SrTiO_3$ substrates over the temperature range of 600-900 °C.

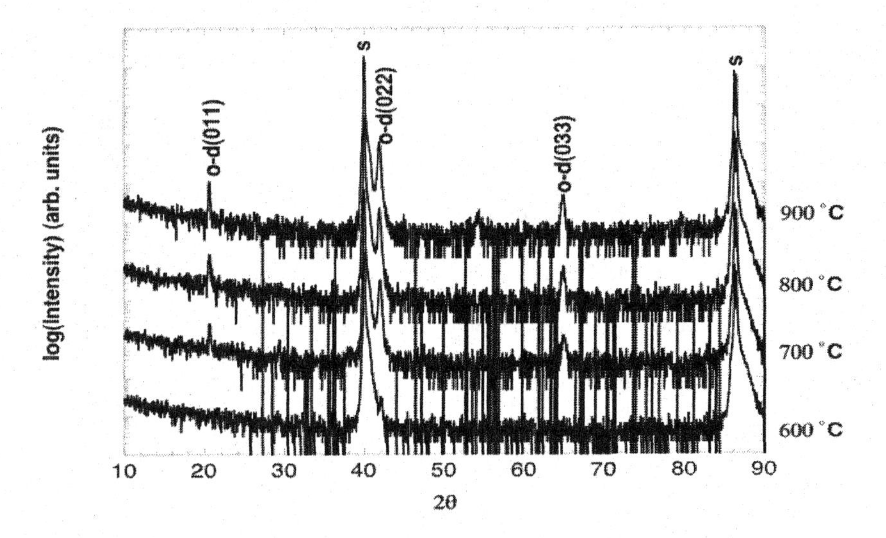

Figure 4. Normal θ–2θ scans of orthorhombic $DyMnO_3$ films grown on $SrTiO_3$(111)

At 600 °C, the $DyMnO_3$ film yields only a very weak diffraction peak (the substrate peaks are marked s), while at higher temperatures we observe good quality, (011) oriented $DyMnO_3$ [peaks labeled o-d(0*kk*)]. The (011) orientation is one of the close-packed planes in the distorted perovskite cell of orthorhombic $DyMnO_3$, and is structurally similar to $SrTiO_3$(111). This orientation was not observed in the $DyMnO_3$ film deposited upon MgO(111) at 900°C, and is stabilized because it helps minimize the interfacial and strain energy terms in Eq. 1. $SrTiO_3$ has a lattice parameter of 3.905 Å, and a [110] distance of 5.52 Å; the periodic repeat distance along the relevant direction in the (011) plane of orthorhombic $DyMnO_3$, which is a <111> type direction, is 5.45 Å. This gives a mismatch of about 1 % and an epitaxial relationship described as: $(011)_{DyMnO3} \parallel (111)_{SrTiO3} \mid [11\bar{1}]_{DyMnO3} \parallel [1\bar{1}0]_{SrTiO3}$.

We can now consider the different terms in the free energy expression, Eq. 1. The G_v term is approximated by the appropriate value given in Figure 1, where the hexagonal phase is predicted to be more stable than the orthorhombic phase by •2 kJ/mol. On MgO, neither phase forms a coherent interface, implying that the interfacial energy will be elevated for both cases. It is clear, based on the observed epitaxial arrangement, that the hexagonal phase indeed forms a semi-coherent interface, which will lead to a smaller $\gamma_{Film/Subs}$ term for it than for the orthorhombic phase. In fact, the observed perovskite peaks at 900°C arise from orientations that do not have a three fold axis perpendicular to them, leading to increased incoherency at the interface. The (001) plane of the hexagonal structure is a natural cleavage plane, implying that the surface energy term should also be relatively small for the observed orientation. Finally, the strain energy term is expected to be low for semi-coherent and incoherent interfaces as well. As such, the observation of hexagonal $DyMnO_3$ at 800 °C may arise from either the absolute thermodynamic preference of the bulk hexagonal phase at this temperature, or it may arise from a minimization of the interfacial energies during nucleation and growth. Hence, it is impossible to comment on the absolute stability of $DyMnO_3$. What is clear, however, is that the orthorhombic phase begins to compete with the hexagonal phase at elevated temperatures on this surface. The observation of the orthorhombic phase at these higher temperatures definitely argues for a thermodynamic instability of the hexagonal phase at increased temperatures, which we believe to be an indication that the thermodynamically stable phase at 900 °C is the orthorhombic variant (perhaps arising from defects). In a similar fashion, the perovskite polymorph is stabilized on the perovskite substrate because interfacial and strain energies are minimized by forming a coherent film/substrate interface, whether or not the absolute thermodynamic preference is for the perovskite or hexagonal phase. The hexagonal film must grow in a semicoherent or incoherent fashion, which would greatly increase the free energy of formation of the hexagonal phase compared to the perovskite. Indeed, we see peaks only from the orthorhombic $DyMnO_3$ phase over the entire temperature range.

Although the absolute thermodynamic preference is unknown for $DyMnO_3$, it is well accepted that the perovskite phase is stable for $GdMnO_3$. As such, films of $GdMnO_3$ were deposited on MgO(111) and $SrTiO_3$(111) substrates to test the stability of this compound. Figure 5 shows the XRD patterns of these films deposited at 800°C and 5 mTorr O_2.

From Figure 5, we can see that $GdMnO_3$ grows as a single phase orthorhombic material on a perovskite substrate $SrTiO_3$(111), in a fashion similar to the growth of $DyMnO_3$. Only peaks from the substrate (marked **s**) and the (011)-oriented orthorhombic phase (marked o-g(o*kk*)) are observed. For $DyMnO_3$ on MgO(111) (MgO peaks are marked **m**), on the other hand, a mixture of the two phases is observed. The hexagonal phase is labeled as h-g(00*l*) and the orthorhombic as o-g(121). Keeping in mind that the stable form for $GdMnO_3$ in the bulk is the orthorhombic structure, it is interesting to observe the hexagonal structure in its metastable form on MgO(111), with which it only forms a semicoherent interface. Based on the approximations used to create Figure 1, we can see that the difference in free formation energies between the two phases is about 5 kJ/mol. Gorbenko *et al*[20] have argued that if the difference in free formation energies is about 10 kJ/mol., then the metastable phase can be stabilized. Bosak *et al*[12] found a mixture of the two phases for $GdMnO_3$ deposited on YSZ(111) using chemical vapor deposition.

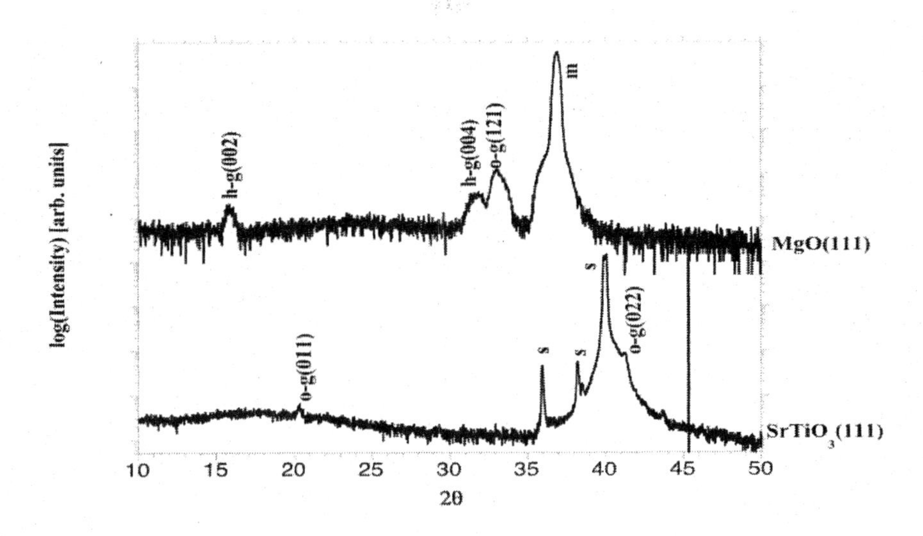

Figure 5. XRD pattern of the $GdMnO_3$ film deposited on plain $SrTiO_3$ and MgO(111) substrates.

In that case, they believed that they exceeded the critical thickness and that a phase transformation occurred to the stable phase. This is similar to our result on MgO(111) substrates where we find that the $GdMnO_3$ film grows as a biphasic mixture. The implication here is that the values of strain and surface energies are not enough to overcome the natural preference for the orthorhombic phase. These observations also argue that the semi-coherent interface between the hexagonal structure and MgO(111) helps to partially stabilize the metastable phase for $GdMnO_3$. This observation further illustrates that the absolute thermodynamic preference cannot be determined for $DyMnO_3$ in these experiments, because we see metastable hexagonal phases on MgO(111). These results for $DyMnO_3$ and $GdMnO_3$ films grown on MgO(111) in the temperature range of 600-900 °C are summarized in Table 1 and compared to $YMnO_3$. Most importantly, if a substrate that had a coherent interface with the hexagonal $GdMnO_3$ phase was used, it seems that this material could be synthesized phase pure in the hexagonal form.

Temperature (°C)	$YMnO_3$	$DyMnO_3$	$GdMnO_3$
900	Single phase hexagonal	Hexagonal + Orthorhombic	Hexagonal + Orthorhombic
800	Single phase hexagonal	Single phase hexagonal	Hexagonal + Orthorhombic
700	Single phase hexagonal	Single phase hexagonal	Hexagonal + Orthorhombic
600	Not Crystallized	Not Crystallized	Not Crystallized

Table 1 Summary of Phase selection across the *RE*MnO_3 series as a function of temperature on MgO(111) substrates.

CONCLUSION

We have succeeded in growing phase-pure orthorhombic $DyMnO_3$ films on $SrTiO_3(111)$ substrates. The hexagonal form grows on MgO(111) substrates until a certain temperature after which decomposition to the orthorhombic polymorph is initiated. The phase pure polymorphs have been epitaxially stabilized by using substrates with low energy interfaces. The orthorhombic form grows with the (011) orientation and the hexagonal orientation grows with the (001) orientation, both of which are typical in this family of materials. We have also succeeded in producing the hexagonal ferroelectric polymorph of a larger rare earth manganite, $GdMnO_3$. Still, we have not achieved phase-pure hexagonal material growth, and the possibility of modifying the substrate surface to produce single phase films is under investigation. The ferroelectric properties of these films are still to be investigated. While there is no limit to the production of the metastable orthorhombic variant in the case of the smaller rare earth manganites, as all the compounds in the series have been obtained in the metastable form, clear limits to the stability of synthesizing the metastable hexagonal structure for the larger rare earth manganites as thin films are observed. We are investigating these materials to understand better thin film synthetic approaches.

REFERENCES

1. H.L. Yakel, W.C. Koehler, E.F. Bertaut et al., "On the Crystal Structures of the Manganese (III) Trioxides of the Heavy Lanthanides and Yttrium.," *Acta Crystallographica* **16**, 957 (1963).
2. T. Yoshimura, N. Fujimura, and T. Ito, "Ferroelectric properties of c-oriented $YMnO_3$ films deposited on Si substrates," *Appl. Phys. Lett.* **73** (3), 414 (1998).
3. N. Fujimura, T. Ishida, T. Yoshimura et al., "Epitaxially grown $YMnO_3$ film: New candidate for nonvolatile memory devices," *Applied Physics Letters* **69** (7), 1011 (1996).
4. F. Moussa, M. Hennion, J. Rodriguez-Carvajal et al., "Spin waves in the antiferromagnet perovskite $LaMnO_3$: A neutron-scattering study," *Physical Review B* **54** (21), 15 149 (1996).
5. M.A. Gilleo, "Crystallographic studies of perovskite-like compounds. III. $La(M_x,Mn_{1-x})O_3$ with M = Co, Fe and Cr.," *Acta Crystallographica* **10**, 161 (1957).
6. R.v. Helmolt, J. Wecker, B. Holzapfet et al., "Giant negative magnetoresistance in perovskitelike $La_{2/3}Ba_{1/3}MnO_x$ ferromagnetic films," *Physical Review Letters* **71**, 2331 (1993).
7. N.Q. Minh, "Ceramic Fuel Cells," *Journal of the American Ceramic Society* **76**, 563 (1993).
8. S. Geller and E.A. Wood, "Crystallographic Studies of Perovskite-Like Compounds. I. Rare Earth Orthoferrites and $YFeO_3$, $YCrO_3$, $YA10_3$," *Acta Crystallographica* **9**, 563 (1956).
9. A. Waintal and J. Chenavas, "High-pressure transformation of the hexagonal form of $MnT'O_3$ (T'=holmium, erbium, thulium, ytterbium, lutetium) into a perovskite form. ," *Material Research Bulletin* **2**, 819 (1967).
10. G. Szabo, These, University of Lyon, 1969.
11. P.A. Salvador, T.-D. Doan, B. Mercey et al., "Stabilization of $YMnO_3$ in a Perovskite Structure as a Thin Film," *Chem. Mater.* **10**, 2592-2595 (1998).

[12] A.A. Bosak, C. Dubordieu, J.P. Senateur et al., "Hexagonal Rare earth (R = Eu - Dy) manganites: XRD study of epitaxially stabilized films.," *Crystal engineering* **5**, 355 (2002).

[13] T.-D. Doan, C. Abramowski, and P.A. Salvador,"Stability and structural characterization of epitaxial $NdNiO_3$ films grown by pulsed laser deposition.," presented at the *Materials Research Society Symposium Proceedings*, (2001), 658(Solid-State Chemistry of Inorganic Materials III), GG3.27.21

[14] A.J. Francis and P.A. Salvador,"Synthesis, structures, and physical properties of yttrium-doped strontium manganese oxide films. ," presented at the *Materials Research Society Symposium Proceedings* (2002), 718(Perovskite Materials), 163

[15] B. Mercey, P.A. Salvador, W. Prellier et al., "Thin film deposition: a novel synthetic route to new materials.," *Journal of Materials Chemistry* **9** (1), 233 (1999).

[16] A.J. Francis, A. Bagal, and P.A. Salvador, in *Innovative Processing and Synthesis of Ceramics, Glasses and Composites IV*, edited by N. Bansal (The American Ceramic Society, Inc., 2000), p. 565.

[17] R.D. Shannon and C.T. Prewitt, "Effective Ionic Radii in Oxides and Fluorides," *Acta Crystallographica B* **25**, 925 (1969).

[18] K. Kamata, T. Nakajima, and T. Nakamura, "Thermogravimetric study of rare earth manganites $AMNO_3$ (A=samarium, dysprosium, yttrium, erbium, ytterbium) at 1200 °C.," *Mat. Res. Bull.* **14** (8), 1007-1012 (1979).

[19] H. Satoh, J.-i. Iwasaki, K. Kawase et al., "High temperature enthalpies and heat capacities of $YbMnO_3$ and $YMnO_3$," *Journal of Alloys and Compounds* **268** (1-2), 42 (1998).

[20] O.Y. Gorbenko, S.V. Samoilenkov, I.E. Graboy et al., "Epitaxial stabilization of oxides in thin films.," *Chemistry of Materials* **14**, 4026 (2002).

MULTI-FERROIC $BiFeO_3$ FILMS PREPARED BY LIQUID PHASE EPITAXY AND SOL-GEL METHODS

X. Qi*, P.S. Roberts, N.D. Mathur, J.S. Lee[1], S. Foltyn[1], Q.X. Jia[1], J.L. MacManus-Driscoll

Department of Materials Science and Metallurgy, University of Cambridge, Pembroke Street, Cambridge CB2 3QZ, UK. [1]Superconductivity Technology Center, Los Alamos National Laboratory, Los Alamos, NM 87545, USA

ABSTRACT

Liquid phase epitaxy (LPE) and sol-gel methods have been used to grow $BiFeO_3$ thin films on $SrTiO_3$ and $LaAlO_3$ single crystal substrates. XRD pole figures showed that the films were bi-axially aligned, with an in plane texture of about 2° and out-plane texture of 0.6°. The LPE films grown on (001) $SrTiO_3$ had a 4-fold symmetry along the c-axis, indicating they had a heterostructure different from the bulk crystal. Both piezoresponse force microscopy and the ferroelectric measurements showed that the films were ferroelectric. Vibrating sample magnetometry showed that the LPE grown films were ferromagnetic.

1. INTRODUCTION

$BiFeO_3$ has a rhombohedrally distorted peroskite structure with space group R3m and a=0.396nm, α=89.5° [1, 2]. It belongs to a special class of material, in which both electric and magnetic ordering can coexist. Such materials are rare in nature, but recently there has been a growing interest in them because of many potential applications including new memory and sensor devices. Bulk $BiFeO_3$ material was reported to be antiferromagnetic below the Neel temperature of 647 K [3,4] and ferroelectric with a high Curie temperature of 1043 K[3, 4]. Thin film form of $BiFeO_3$ sometimes showed different structures and properties, which were probably induced by hetero epitaxial growth or incorporation of impurities [5]. So far, most of the studies have been carried out using bulk polycrystalline samples. Therefore, it is of great interest to carry out more studies on epitaxial thin film growths. In this paper, we report on the epitaxial growth of $BiFeO_3$ thin films by the sol-gel and liquid phase epitaxy (LPE) methods. Sol-gel is a versatile and quick method, which is capable of growing thin films on a wide range of substrates. So it is a good method to produce different heterostructures on various substrates. LPE is a process similar to single crystal growth from high temperature solution. Therefore it is capable of producing high structural perfection, high purity films.

2. EXPERIMENTS

Sol-gels were prepared using both citrate and acetate precursors. In the citric route, the Bi_2O_3 oxide powder was first dissolved in diluted nitric acid solution and then mixed with the same amount of the iron (III) citrate water solution. The concentration of the metal ions in the solution was 0.02M. 30g of citric acid was dissolved in 50ml distilled water and then mixed with 100ml of the above solution and finally, 30ml of ethylene glycol was added. The mixed solution was heated at 70-80 °C with vigorous stirring to evaporate water and after about 6 hours, a viscous yellow gel solution was formed. In the acetic route, the starting materials were bismuth (III) acetate and iron (III) acetylacetonate, which were first dissolved in acetic acid and acetone,

* Corresponding author, e-mail xq204@cam.ac.uk

respectively. Then, the two solutions were mixed and evaporated at 60 °C for 3-4 hours to form a viscous gel solution. Ethylene diamine was added to adjust the pH value at 3.5-4.0 during the heating. Coatings of the gel solutions on (001) $SrTiO_3$ (STO) and (001) $LaAlO_3$ (LAO) substrates was made through fast spinning of 4000 rpm. The samples were dried at temperatures between 100-400 °C for several hours and then fired at 780 °C in air for 1 to 12 hours.

LPE growth of $BiFeO_3$ was carried out from the 85% Bi_2O_3+15% Fe_2O_3 flux contained in an alumina crucible. In order to avoid the high temperature liquid being contaminated by the crucible, the inside surface of the crucible was coated with a thin layer of Fe_2O_3, which was made by painting on iron citrate gel and then sintering at 1200 °C for 24hr. The crucible, charged with well-mixed Bi_2O_3+Fe_2O_3 powder, was heated up to 830 °C to form a single liquid state. After soaking for about 6 hrs, the temperature was quickly reduced to 820 °C, at which the (001) STO or $SrRuO_3$ buffered STO substrates were brought down slowly and dipped into the liquid. The substrates were then rotated at 60 rpm to circulate the high temperature solution and after growing for a desired period of time, they were pulled clear from liquid and slowly brought up to the room temperature in about 30 min.

The grown films were characterised by x-ray diffraction (XRD), scanning electron microscopy (SEM), energy disperse spectroscopy (EDS), vibrating sample magnetometry (VSM), piezoresponse force microscopy (PFM) and standard ferroelectric measurements for determining the electric hysteresis loops.

3. RESULTS AND DICUSSIONS

Pure and oriented thin films of $BiFeO_3$ could be grown on the (001) STO and LAO substrates from both acetic and citric sol-gels. A typical XRD θ -2θ scan was showed in Fig.1, which was recorded using a thin film grown on (001) LAO and showed no other reflection lines except for the (00*l*) reflections of $BiFeO_3$ and LAO. Such good c-texture could only be maintained in the film thickness up to about 200-300nm. Other orientations started to occur in thicker films, as indicated in both XRD and surface SEM. Acetic sol-gels were found to have a better wetting property than the citric sol-gels. Hence, uniform coating was much easier to achieve. However, the final composition and surface morphology of the acetic sol-gel films were very sensitive to the drying and sintering procedure. The firing temperatures and times had to be controlled precisely in order to avoid preferential evaporations of bismuth containing compounds, which resulted in a porous surface as well as the formation of an iron-rich phase $Bi_2Fe_4O_9$. Sintering at lower temperatures (500-600 °C) was required before final sintering at higher temperatures. In contrast, citric sol-gel films

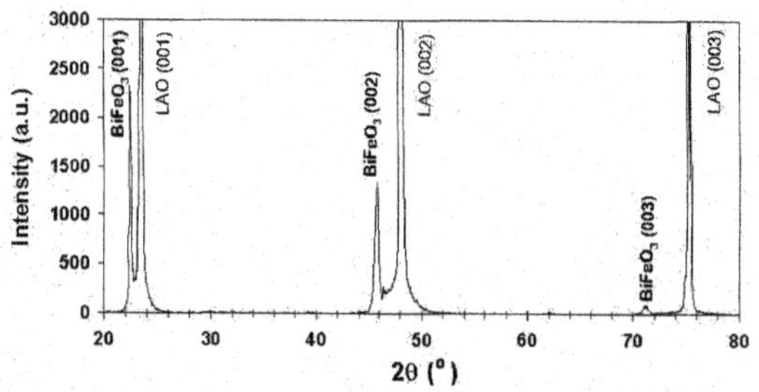

Fig.1 XRD θ-2θ of sol-gel film grown on (001) LAO

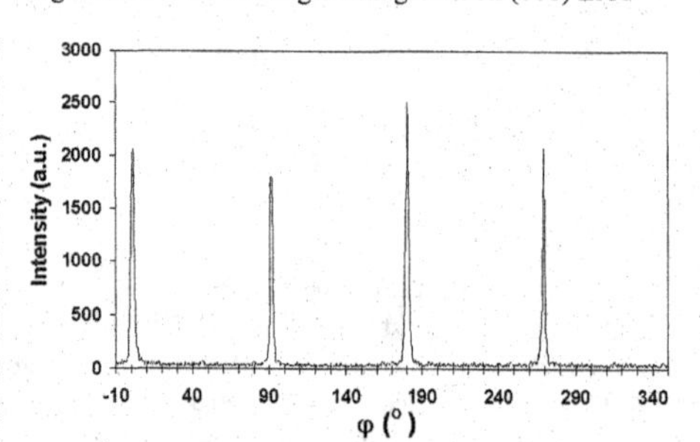

Fig.2 XRD φ-scan of LPE film grown on (001) STO

were more stable and could be sintered at 780 °C for 12 hours without forming notable amount of other phases.

Sol-gel films were generally more porous compared to the LPE-grown films, which were highly textured as well. XRD θ-2θ scan of the LPE films grown on the (001) STO substrates showed a similar characteristic as shown in Fig.1, which recorded only the (00*l*) reflections of $BiFeO_3$ and the STO substrate, indicating the films were pure and oriented along the c-axis. A typical ϕ-scan of the (103) reflection of LPE films is given in Fig.2, and shows that the films had a 4-fold symmetry along c-axis, suggesting they were either tetragonal or cubic, which were different from the bulk crystal. The films grown on (001) STO were highly bi-axially aligned, with a full width half maximum (FWHM) of about 2° for the (103) ϕ-scan and 0.6° for the rocking curve of the (001) reflection. LPE was unable to grow $BiFeO_3$ films on the (001) LAO substrates due to a large lattice misfit of about 5%. The achievable supercooling in the Bi_2O_3:Fe_2O_3 high temperature solution before spontaneous nucleation occurred was about 20-30 °C. Such a supercooling was not large enough to produce sufficient supersaturation for the nucleation on the LAO substrates.

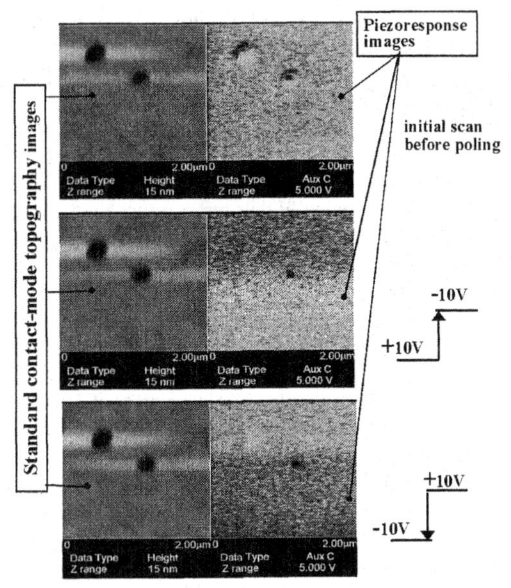

Fig.3 PFM images of LPE film grown on Nb:STO, showing appearance of polarizations after applying electric fields.

Ferroelectric measurements were first carried out by the PFM imaging using a LPE film grown on the 1 at% Nb^{5+} doped, conducting STO substrate. The left images shown in Fig.3 are the standard topographic images and the right ones are the piezoresponse images, which show the occurrence of electric polarization after applying the bias, indicating the film was ferroelectric. The LPE films were found to have a low resistance, which was approaching to the same magnitude as the Nb^{5+} doped STO substrates. Therefore, the ferroelectric hysterisis measurements were carried out using the films grown on the $SrRuO_3$ buffered STO substrates. The conductive $SrRuO_3$ layer was used as the bottom electrode for applying electric field across the films from the top, which was evaporated with a layer of gold. The results confirmed the ferroelectric behaviour of the films and the hysteresis loops are shown in Fig.4, which also shows a large leakage current at the voltages higher than 6V. The hysteresis loops in Fig.4

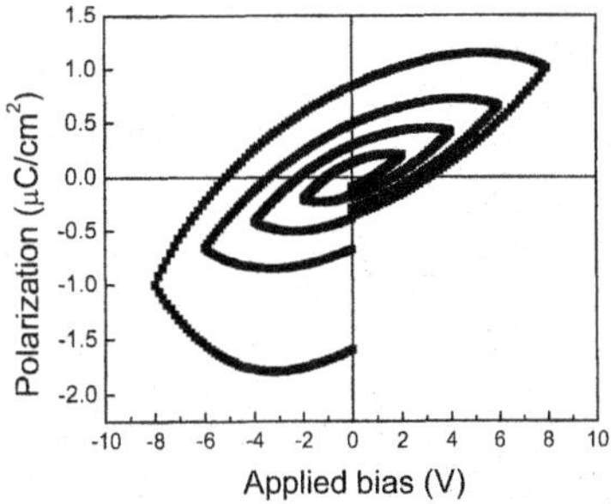

Fig.4 Ferroelectric hysteresis loops of LPE film grown on (001) $SrRuO_3$/STO

could also result from a capacitor with the leakage resistance, but the hysteresis loop measurements in conjunction with the previous PFM measurements suggest a true remnant polarization state. The low resistance of the LPE films might result from a relatively high level of Fe^{2+} ions in the films, since they were grown from the melt at a much higher temperature than in other growth methods. Considerable amount of divalent irons has been found in other ferrite single crystals grown from high temperature melts [6]. Fe^{2+} has been proposed to induce mobile carriers and hence low electric resistance in many ferrites (e.g. MnZn ferrite) [7, 8] and other dielectric oxides [9]. Actually, single crystals of $BiFeO_3$, which were grown from the melt under very similar conditions as LPE (but for much longer time) [10], were too conductive to measure the dielectric hysteresis at room temperature, and the samples had to be cooled in liquid nitrogen [10].

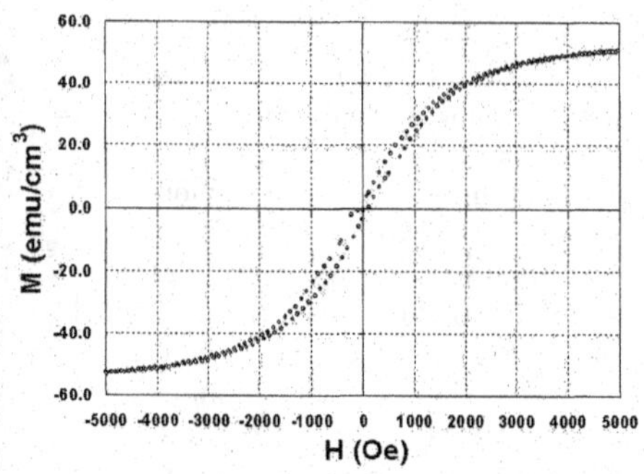

Fig.5 M-H loop of LPE film grown on STO

The magnetic properties of the films were characterized by VSM at room temperature. There was a small linear increase of magnetization with applied field for the sol-gel films grown both on the STO and LAO substrates, indicating the films were either antiferromagnetic or paramagnetic. However, the LPE films were observed to be ferromagnetic and a magnetic hysteresis loop measured with the field perpendicular to the film is shown in Fig.5. The origin of the ferromagnetism in the LPE films is unclear. It possibly came from a relatively high level of Fe^{2+} ions, as discussed above. The incorporation of larger Fe^{2+} ions (15% larger than Fe^{3+}) would introduce a large local structural distortion, as well as possible double exchange interaction between 2+ and 3+ ions through the oxygen. However, more experiments are needed to support this assumption.

4. CONCLUDING REMARKS

Highly textured $BiFeO_3$ thin films have been grown on the (001) STO and LAO single crystal substrates by LPE and sol-gel methods, with an in-plane texture of 2° and out-plane texture of 0.6°. LPE films grown on STO substrates showed a 4-fold symmetry along the c-axis, indicating they had a heterostructure different from the bulk crystal. Both PFM imaging and standard ferroelectric measurements showed that the LPE films were ferroelectric. VSM showed that the LPE films were ferromagnetic, which possibly resulted from a relatively high level of Fe^{2+} in the films.

REFERENCES

[1] A.G. Tutov, "The space group and some electric and magnetic properties of bismuth ferrite", FTVTA, **11**, 2681 (1969)

[2] G.D. Achenbach, R. Gerson, W.J. James, "The atomic structure of $BiFeO_3$", J. Am. Ceram. Soc. **50**, 437 (1967)

[3] V.A. Murashav, D.N. Rakov, V.M. Ionov, I.S. Dubenko, Y.U. Titov, Ferroelectrics **162**,11(1994).

[4] Yu. F. Popov, A.M. Kadomtseva, G.P. Vorobev, A.K. Zvezdin, Ferroelectrics **162**, 135 (1994).

[5] J. Wang, J.B. Neaton, H. Zheng, V. Nagarajan, S.B. Ogale, B. Liu, D. Viehland, V. Vaithyanathan, D.G. Schlom, U.V. Waghmare, N.A. Spaldin, K.M. Rabe, M.Wutting, R. Ramesh; "Epitaxial $BiFeO_3$ Multiferroic Thin Film Heterostructures", Science, **299**, 1719 (2003).
[6] G. Winkler, "Crystallography, Chemistry, and Technology of Ferrites", in Magnetic Properties of Materials, Inter-University Electronics Series, VOL13, Edited by J. Smit, 1971, McGraw Hill, pp20-63.
[7] J. Verweel, "Ferrites at Radio Frequencies", in Magnetic Properties of Materials, Inter-University Electronics Series, VOL13, Edited by J. Smit, 1971, McGraw Hill, pp64-107.
[8] S.S. Bellad and B.K. Chougule, "Composition and frequency dependent dielectric properties of Li–Mg–Ti ferrites", Mater. Chem. Phys. **66**, 58(2000)
[9] I. Nee, M. Muller, K. Buse, and E. Kratzig, "Role of iron in lithium-niobate crystals for the dark-storage time of holograms", J. Appl. Phys., **88**(7), 4282(2000)
[10] J.R. Teague, R. Gerson and W.J. James, "Dielectric Hysteresis in Single Crystak $BiFeO_3$", Sol. Stat. Comm., **8**, 1073(1970)

EFFECT OF ARGON ADDITION DURING ECR MODE NUCLEATION OF DIAMOND FILMS GROWN BY MPCVD

Vidhya Sagar Jayaseelan, V. Shanov and R. N. Singh
401 Rhodes Hall,
Department of Chemical and Materials Engineering,
University of Cincinnati,
Cincinnati, OH 45221-0012

ABSTRACT

Diamond films were deposited on (100) silicon wafers in a two step process of low pressure Electron Cyclotron Resonance (ECR) mode nucleation followed by high pressure microwave plasma enhanced chemical vapor deposition (MPCVD). The ECR nucleation process at 4.67 Pa in a hydrogen-argon-methane plasma led to the formation of nuclei of carbon based material of 20-50 nm size with high number density in addition to large transparent grains of about 5-50 μm size. Optical emission spectroscopy indicated higher degree of ionization with increasing argon during the nucleation stage. Raman spectra of the large grains did not show the characteristic diamond peak. Further deposition at 4000 Pa on the nucleated substrates led to the formation of discontinuous diamond films with well pronounced (100) facets. Raman spectra of these films showed the first order phonon peak of diamond at 1332 cm^{-1} as well as a broad peak around 1555 cm^{-1} indicating presence of some sp^2 carbon content. There was no orientation relationship between the cubic axes of the silicon substrate and those of the (100) facets on the films ruling out diamond epitaxy. Increasing the argon concentration in the hydrogen-methane plasma during the nucleation step had the effect of decreasing the nucleation density of diamond. Addition of argon during the ECR stage did not generate any specific advantage to the diamond films grown by the MPCVD process although the stability of the plasma and the degree of ionization were higher.

INTRODUCTION

Diamond is a very promising material for electronic and optical applications for high frequency and high power devices because of its large band-gap, extreme mechanical properties, high thermal conductivity and chemical inertness. Though this has been demonstrated [1], a lot of new ground has to be covered especially in the areas of epitaxial growth of semiconductor quality diamond, its doping and micro-fabrication techniques before commercial application can be achieved. Recent success in the growth of epitaxial diamond has further intensified efforts in these directions to exploit these materials as semiconductors [2,3,4].

When a constant magnetic field is applied to microwave plasma, the free electrons in the plasma experience Lorentz force leading to their revolution around the magnetic field lines in spiral orbits. The frequency of revolution in these spiral orbits is dependent only on the applied magnetic field. If the magnetic field strength is adjusted such that this frequency matches the incoming microwave frequency, the oscillating electrons resonate with the microwave leading to maximization of power absorption by the plasma. This is known as the electron cyclotron resonance (ECR) condition, which leads to generation of high intensity plasmas with the ionization density several orders of magnitude higher than those achieved without the magnetic field especially at lower pressures. It also results in higher electron temperature and subsequently greater ion bombardment across the larger Debye sheath voltage drop of the substrate. Ion bombardment by biasing the substrate has been an important feature of growth of epitaxial diamond films [2,3]. The alternative to biasing, to obtain oriented films, is the use of an ECR plasma nucleation step [5,6] to obtain more uniform ion bombardment with significantly greater electron temperature.

The molecular species, CH_3, C_2H_2 and C_2 are thought to be the main carriers of carbon formed from methane that control the formation of diamond by MPCVD. While CH_3 and C_2H_2 [7] are believed to play a major role at high hydrogen concentrations, mechanisms based on C_2 are said to dominate at high Ar concentrations [8,9]. Hydrogen has been the primary carrier gas for formation of diamond by CVD processes. It is believed that H_2 addition helps diamond formation by stabilization of sp^3 phase through elimination of dangling bonds, selectively etching the graphite and amorphous phases of carbon and by generating the high pressure high temperature conditions required for diamond formation in the phase diagram by proton (H^+) bombardment [7]. Because of its higher ionization cross-section, addition of Argon to the CH_4-H_2 system has helped in formation of a more stable and high density plasma and enables deposition at higher growth rates [8,9]. It has also been found to decrease the grain size due to secondary nucleation and reduce diamond quality significantly.

In these experiments Argon was added to an ECR mode Hydrogen-Methane plasma nucleation step and the effect on diamond films deposited was studied. Since Argon increases ionization along with greater ion bombardment, it could be expected to improve orientation relationship between the Silicon substrate along with increasing the growth rates which is a significant problem especially at low pressure conditions as used under the ECR mode.

EXPERIMENTAL METHOD

P-type Silicon (100) wafers diced into squares were utilized as substrates for the study. Silicon samples were used in the commercially available polished condition and also after ultrasonic activation for 2 hours in a 1-2 μm diamond slurry in ethanol. All the silicon samples were treated in dilute 2% HF solution for 10 minutes to remove the nascent oxide layer.

The deposition process was carried out in an ASTEX AX5100- MPCVD system shown in Fig.1. The details of this system have been described elsewhere [10]. The microwave power source has the standard industrial frequency of 2.45 GHz. The magnetic field strength at which ECR is obtained for this frequency is 875G. DC current of 180A and 120A were passed through the top chamber's entry and exit electromagnets, generating ECR condition between the magnets.

The deposition process comprised of 2 steps- ECR mode nucleation and deposition. During the ECR mode nucleation, the substrate holder was moved down into the bottom chamber and methane was introduced directly into the bottom chamber above the substrate while all the other gases were introduced from the top chamber. The pressure was maintained at 4.67 Pa and the temperature of the substrate was maintained at 700°C. The plasma gas mixture comprised of methane (33.3%) and varying amounts of hydrogen and argon. The microwave power input was maintained at 300W and the reflected power was typically kept below 20W. Hydrogen plasma etching was carried out for 15 minutes to remove organic or volatile impurities and native oxide before nucleation and to etch away sp^2 carbon after nucleation. Nucleation was done for 5 hours. During nucleation optical emission from the plasma discharge was analyzed with an Ocean-optics HR2000CG-UV-NIR composite grating spectrometer.

After nucleation, the second step involved deposition at a pressure of 4000 Pa in the top chamber with the ECR magnets turned off. The temperature was maintained at 700°C. After 15 minutes of etching in 500W hydrogen plasma, the diamond films were deposited in a 900W, 1% methane in hydrogen plasma for 3 hours, followed by 15 minutes of hydrogen etching again to avoid deposition of sp^2 carbon on the films. All gases were introduced from the top chamber. Conditions for nucleation and growth have been summarized in Table-I.

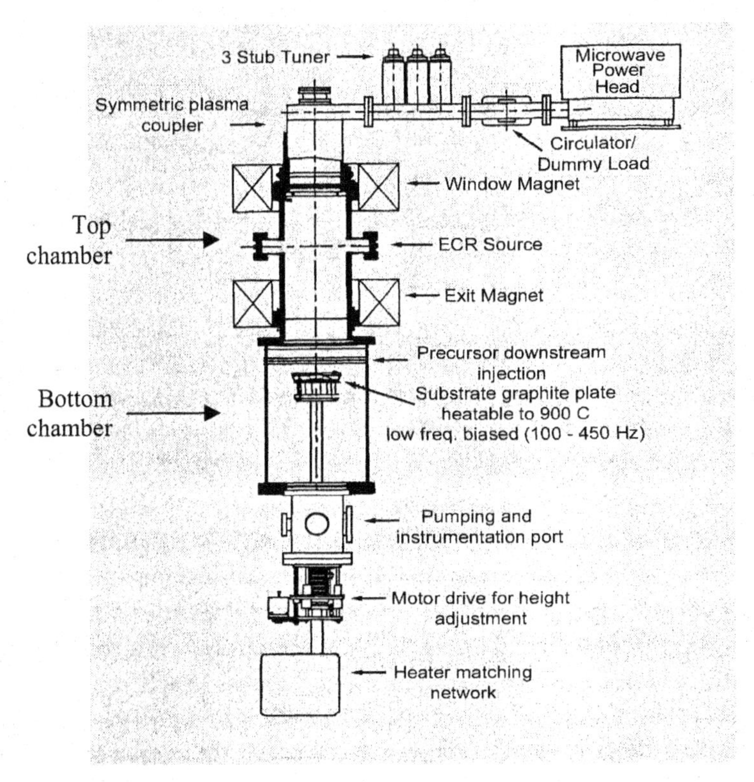

Fig.1: Schematic of the MPECVD system showing the electromagnets used to generate the ECR condition, and the bottom and top chambers, where the substrates were located during nucleation stage and growth stages, respectively

Table-I: Summary of conditions during the nucleation and growth stages of the diamond films with the ECR-MPCVD method.

	Stage 1 – Nucleation			Stage 2 – Deposition
	Sample set 1	Sample set 2	Sample set 3	All samples
CH_4 flow (sccm)	10	10	10	1
H_2 flow (sccm)	20	15	5	99
Ar flow (sccm)	0	5	15	0
Microwave power	300W	300W	300W	900W
ECR	Yes	Yes	Yes	No
Pressure	4.67 Pa	4.67 Pa	4.67 Pa	4000 Pa
Temperature	700 °C	700 °C	700 °C	700°C
Substrate location	Bottom chamber	Bottom chamber	Bottom chamber	Top chamber

The deposited films were observed under an optical microscope and a Hitachi S-4000 Field emission SEM. A Raman spectrometer using T-64000 Jobin Yvon triple monochromator and a 514 nm green Ar laser source was used for analysis of the films. A Nicolet Magna IR 760 spectrometer was used for FTIR spectrometry.

RESULTS AND DISCUSSION

It was observed that as the percentage of argon in the chamber was increased, the plasma could be tuned more easily to a lower value of reflected power. As expected, the optical emission spectra [Fig.2] showed that the Ar peaks (696.7 nm, 750.3 nm, 812.0 nm) have greater intensity with increasing Ar during the ECR mode nucleation. However the increase in intensity of H_α (656.38 nm), C_2 (516.4 nm) and H_β (486.47 nm) with Ar addition indicates greater plasma density and ionization.

After nucleation, the samples were observed under an optical microscope as well as in SEM. Under the optical microscope several large transparent grains were observed [Fig. 3a]. Raman spectra of these grains [Fig.6] had an unknown peak at 492 cm^{-1} in addition to the silicon peak at 520 cm^{-1}. The characteristic diamond peak at 1332cm-1 was absent. Preliminary investigation could not confirm the identity of these grains, though it is suspected to be highly cross-linked or branched alkane polymer. The size of the grains was typically of the order of 5 – 50um. Quite often they were found in clusters. Some of these were faceted with sharp edges while others were smooth edged and appeared glassy or amorphous. The density of these grains was higher on the ultrasonically treated samples and at regions where scratches were present indicating preference for nucleation at high-energy sites.

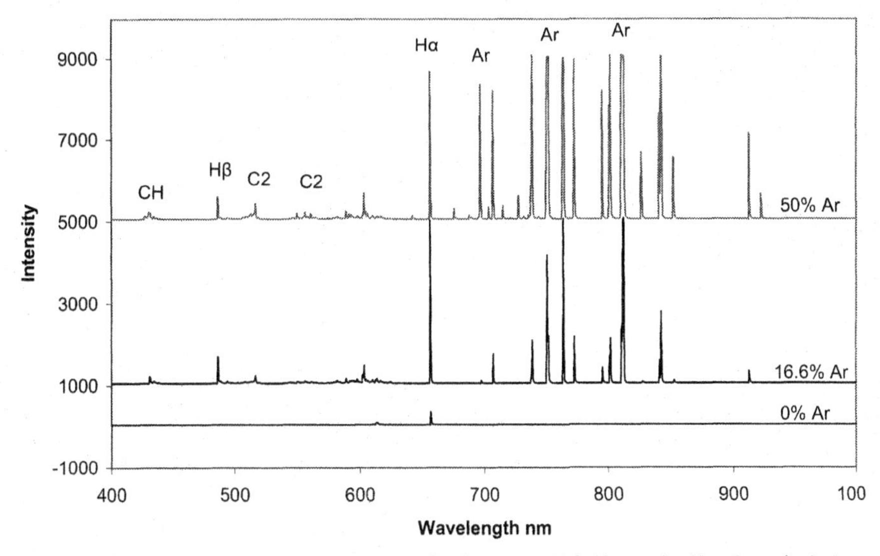

Fig.2: Optical emission spectra of plasma (shifted vertically for clarity) during ECR mode nucleation showing stronger emission lines with increasing Ar concentration not just for Ar peaks and Ar+ but also other species like C2, CH and Hα and Hβ

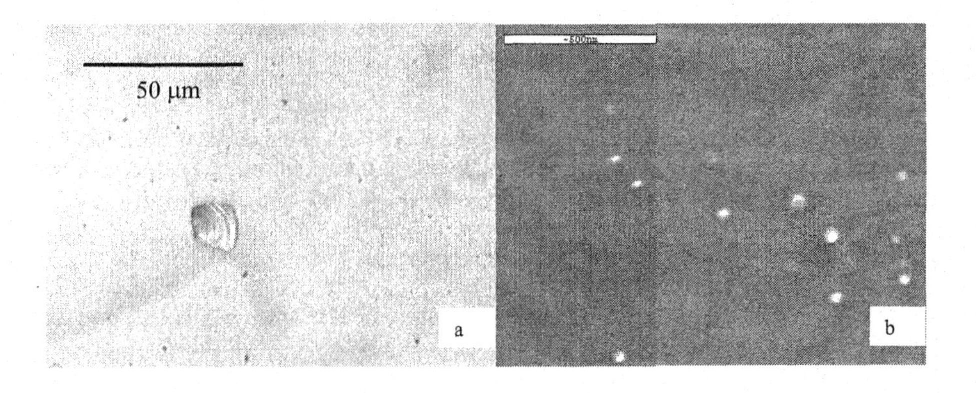

Fig.3: Optical and SEM micrographs of ECR-mode nucleated substrates with no ultrasonic pretreatment. (a) 165X (b) 80000X.

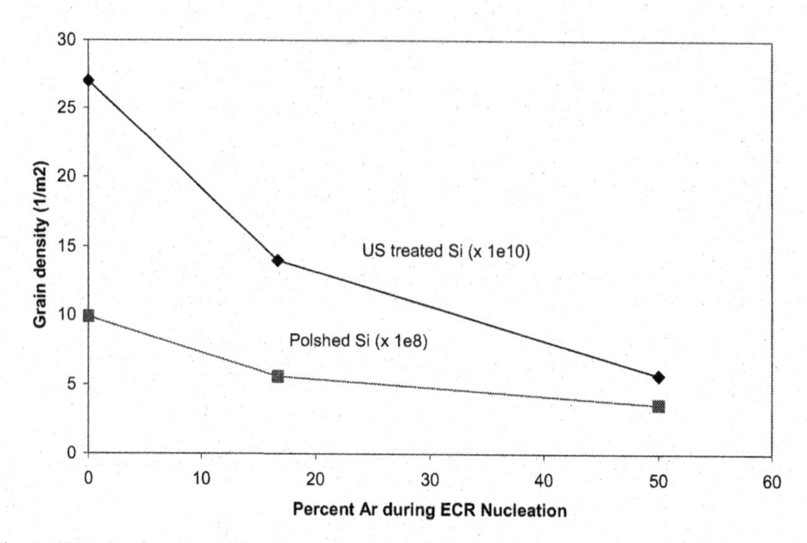

Fig.4: Graph showing the decrease in number density of grains after stage2 (deposition) with increasing Ar concentration during the first stage (nucleation) for both Si substrates ultrasonically treated in diamond slurry as well as untreated ones. Note that the scale of values (given in parenthesis) is different for the two plots

Investigation of the samples under SEM showed the presence of several extremely fine grains [Fig.3b] on the order of 10 – 50 nm size between these large grains [Fig. 3a]. The density of these fine grains was calculated to be of the order of 10^{14} m^{-2}. The samples showed very little contrast in the backscattered electron mode indicating that these are unlikely to be the Gold-Palladium alloy used for metallization of the SEM samples. Similar observations have been made on silicon under similar conditions of growth by other research groups [6]. The three hour deposition step led to the growth of discontinuous diamond films made of polycrystalline grains of the order of 1-2 μm. The number density of grains was significantly higher for the ultrasonically treated samples as compared to the silicon wafers used in the as polished case. Fig.4 shows the effect of increasing Ar concentration during the nucleation stage on the number density of these grains for both the ultrasonically treated samples and the untreated ones. It clearly indicates the decrease in the density with increasing Ar concentration. This could be because the high intensity Ar plasma generated and ion bombardment caused re-sputtering like effects.

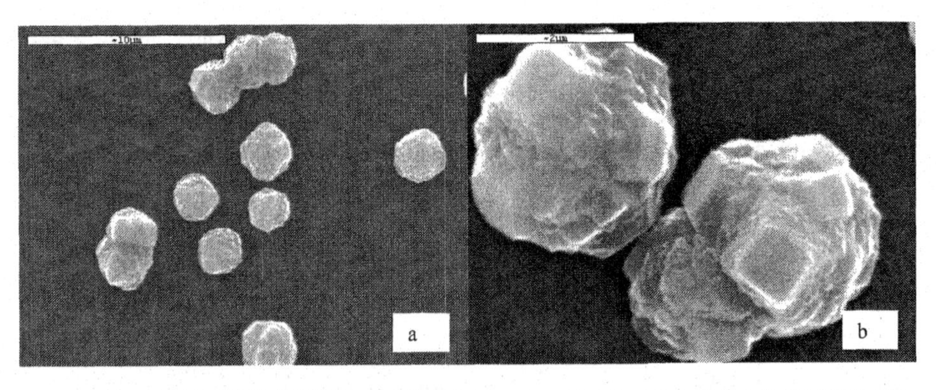

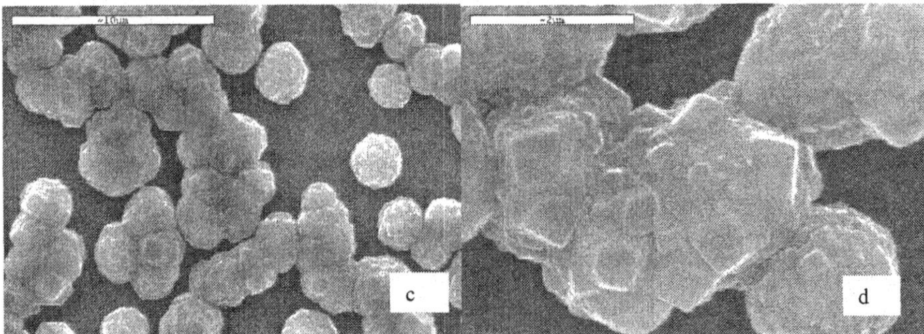

Fig.5: SEM micrographs of diamond grains on silicon substrates after both ECR mode nucleation and MPCVD deposition. (a) untreated substrate 5000X, (b) untreated substrate 20000X, (c) ultrasonically treated in diamond slurry 5000X (d) ultrasonically treated in diamond slurry 20000X. The (100) cubic faceting is observed on several grains and locations within the grain with no apparent relationship with each other.

The morphology of the films indicates growth of cubic crystals with (100) facets growing out of polycrystalline globular diamond aggregates. On several grains many such cubic crystals were observed simultaneously [Fig.5] with no relationship between their individual orthogonal axes ruling out the possibility of epitaxy. The fine nanometer sized nucleated grains were also observed by the group under AFM, though their Raman analysis captured the first order phonon peak of diamond on these nuclei [6]. It is not clear how the nucleation stage affects the diamond film morphology and the velocity of growth of the (100) and (111) planes during the second stage especially since there is no epitaxial relationship with the substrate. Similar cubic morphology had been obtained earlier [6] under two-step deposition processes with ECR mode. It has to be noted that the team however had succeeded later in obtaining preferential orientation of the grains with respect to the Si substrate [11]. Growth rate could not be studied due to the discontinuous nature of the films.

Raman analysis after the deposition process shown in Fig.6, clearly indicates the formation of the diamond through the presence of the first order phonon peak of sp3 phase of carbon at 1332 cm^{-1}. Also the broad peak around 1550cm^{-1} indicates presence of some graphite and amorphous forms of carbon. It has to be remembered that the Raman scattering cross-section due to graphite is 75 times stronger than that due to diamond [12]. The silicon peak at 520cm^{-1} from the substrate is also seen.

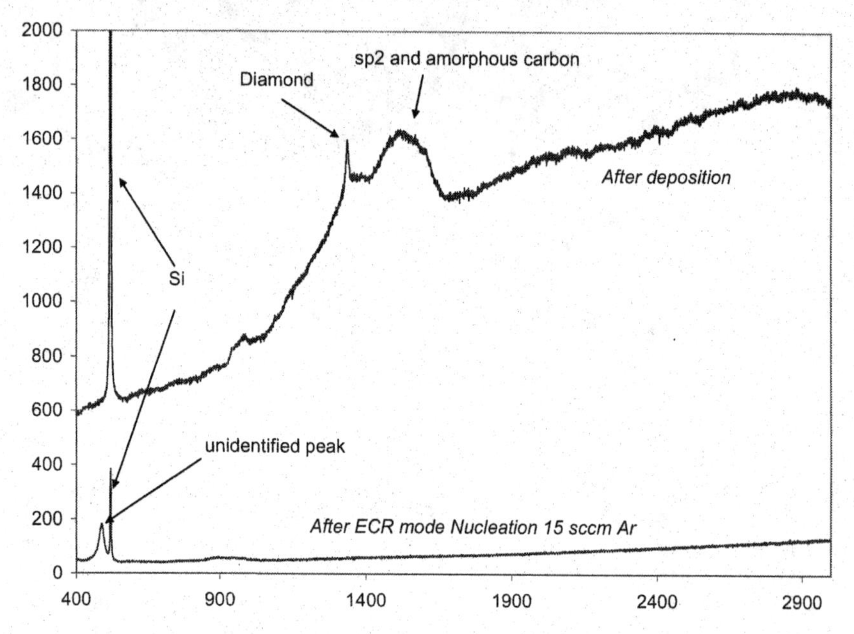

Fig.6: Raman spectra of substrates after ECR mode nucleation with 15sccm Ar flow and a similar substrate after deposition under regular MPCVD conditions. Note the absence of the first order phonon diamond peak (1332 cm^{-1}) in the first case and its presence in the second.

A thick polymer film was coated all over the chamber walls and the microwave inlet quartz window during the ECR deposition mode. The organic nature of the film and the C-C stretching band was identified by FTIR spectroscopy. Although running ECR oxygen plasma at 900W, 100mtorr for 1 hour, could clean the quartz window off, the residue on the walls could not be removed completely since the walls were cooled by chilled water. This is a problem in ECR mode deposition that needs to be tackled perhaps by addition of a small amount of oxygen during deposition or by an Argon gas flow curtain. Also a chamber liner would be useful to help keep the chamber walls free of deposition.

CONCLUSIONS

The ECR nucleation process on (100) silicon led to the formation of nuclei of a carbon based material of 20-50 nm size with high surface density. Raman spectroscopy showed a peak at 492cm^{-1} for large transparent grains of about 5-50 μm size, which were also formed on the substrate. Further deposition at 4000 Pa on the nucleated substrates led to the formation of discontinuous diamond films with well pronounced 100 facets. The sp3 nature of bonding was identified by the first order phonon peak of diamond at 1332 cm^{-1}. Some sp2 material was also present in the film as shown by a broad peak around 1555 cm^{-1}. There was no orientation relationship between the cubic axes of the silicon substrate and those of the (100) facets on the films. Optical emission spectroscopy indicates higher degree of ionization with Ar during the nucleation stage by an increase in the intensity of several peaks including H_{α}, H_{β}, C_2, and CH in addition to Ar. Also addition of Ar during the nucleation decreases the density of grains after the deposition stage for both the ultrasonically treated samples and the untreated ones. There is no effect of Ar introduction, however on the Raman spectra of the deposited films.

Argon was helpful to stabilize the plasma and increase ionization and ECR helped achieve the preferred (100) morphology. Future work along these lines will include study of ECR mode along with bias on the substrates to see if any film modifications can be achieved. Other substrates and addition of process gases like Xe will also be studied.

ACKNOWLEDGEMENTS

The authors would like to thank Dr. Punit Boolchand for help with Raman spectroscopy, Dr. James Boerio for FTIR spectroscopy and Srinivas Subramanian for SEM characterization. This material is based on work supported by the national science foundation under grant no: CMS-0210351. Any opinions, findings, and conclusions or recommendations expressed in this material are those of the author and do not reflect views of the National Science Foundation.

REFERENCES

[1]H.Shiomi, "Diamond Active electronic devices", P579-605, *Diamond films handbook,* Ed. 1, J.Asmussen and D.K.Reinhard, Marcel Dekker Inc., New York, 2002

[2]F. Horrman, H. Y. Peng, Th. Bauer, Q.Li, M. Schreck, Y. Lifshitz, S. T. Lee, B. Stitzker, "Flat epitaxial diamond/Ir(001) interface visualized by high resolution transmission electron microscope." *Surface science,* **513**(2002) 525-529

[3]T. Tsubota, M. Ohta, K. Kusakabe, S. Morooka, M. Watanabe, H. Maeda, " Heteroepitaxial growth of diamond on an Iridium(100) substrate using microwave plasma assisted chemical vapor deposition" *Diamond and related materials,* **9**(2000) 1380-1387

[4]Z. Dai, C. Bednarski-Meinke, R. Loloee and B. Golding "Epitaxial (100) iridium on A-plane sapphire: A system for wafer scale diamond heteroepitaxy." *Applied physics letters,* **Vol. 82**, Number 22, 3847-3849

[5]H. Jeon, C. Lee, A. Hatta, T. Ito, T. Sasaki and A. Hiraki, "Enhancement of diamond nucleation by applying substrate bias in ECR plasma chemical vapor deposition". *Carbon*, **Vol. 36**, No 5-6, pp569-574, 1998

[6]C. Sun W.J. Zhang, C.S. Lee, I. Bello, S.T. Lee, "Nucleation of Diamond films by ECR-enhanced microwave plasma chemical vapor deposition", *Diamond and Related Materials*, **8** (1999) 1410-1413

[7]T.A.Grotjohn, J. Asmussen, "Microwave plasma assisted diamond film deposition" P211-302, *Diamond films handbook*, Ed. 1, J.Asmussen and D.K.Reinhard, Marcel Dekker Inc., New York, 2002

[8]D.Zhou, D.M. Gruen, L.C. Qin, T.G. McCauley and A. R. Krauss "Control of diamond film microstructure by Ar additions to CH4/H2 microwave plasmas." *Journal of applied physics*, **Vol. 84**, Number4 15 Aug 1998, P1981-1989
[9]D. Zhou, T.G.McCauley, L.C. Qin, A.R. Krauss and D.M. Gruen "Synthesis of nanocrystalline diamond thin films from an Ar-CH4 microwave plasma." *Journal of applied physics,* **83**(1), 1January 1998
[10]Astex user manual, PP5-33 in HPM/M Magnetized HPMS plasma source, version 1.6 Woburn, MA, 1992
[11]W.J.Zhang, C.Sun, I.Bello, C.S.Lee and S.T.Lee "A new nucleation method by ECR enhanced microwave enhanced plasma CVD for deposition of (001) oriented diamond films." *Journal of chemical physics*, **Vol 100**, Number 9, 1 Mar 1999.
[12]R.J. Nemanich, I. Bergman, Y.M. Legrice and R.E. Shroder "Raman characterization of diamond film growth", p 741-752, *New diamond science and technology –Conference proceeding,* Ed.1, J.T.Glass, J.E.Butler, R.Roy, MRS, Pittsburgh, 1991

Author Index

Keyword Index

9 781574 981834